MATRICES
AND
TRANSFORMATIONS

Anthony J. Pettofrezzo

Department of Mathematical Sciences and Statistics
Florida Technological University
Orlando, Florida

DOVER PUBLICATIONS, INC.
NEW YORK

Pubished in Canada by General Publishing
Company, Ltd., 30 Lesmill Road, Don Mills,
Toronto, Ontario.
Published in the United Kingdom by Con-
stable and Company, Ltd.

This Dover edition, first published in 1978,
is a republication of the work originally published by Prentice-Hall, Inc., Englewood Cliffs,
New Jersey, in 1966 as part of the Teachers'
Mathematics Reference Series. The "Series
Foreword" has been omitted; but the text is
otherwise unabridged.

International Standard Book Number:
0-486-63634-8
Library of Congress Catalog Card Number:
78-50793

Manufactured in the United States of America
Dover Publications, Inc.
180 Varick Street
New York, N.Y. 10014

To Steven, Paul,
and Donna

Preface

MATRIX ALGEBRA is a significant topic in contemporary mathematics curricula. This book presents the fundamental concepts of matrix algebra, first in an intuitive framework and then in a more formal manner. A variety of interpretations and applications of the elements and operations considered are included. In particular, the use of matrices in the study of transformations of the plane is stressed. The purpose of this book is to familiarize the reader with the role of matrices in abstract algebraic systems, and to illustrate its effective use as a mathematical tool in geometry.

Chapters 1 and 2 include those basic concepts of matrix algebra that are important in the study of physics, statistics, economics, engineering, and mathematics. Matrices are considered as elements of algebra. The concept of a linear transformation of the plane and the use of matrices in discussing such transformations is illustrated in Chapter 3. Some aspects of the algebra of transformations and its relation to the algebra of matrices are included here. Chapter 4 contains material usually not found in an introductory treatment of matrix algebra, including an application of the properties of eigenvalues and eigenvectors to the study of the conics.

The motivations of the concepts presented are included wherever appropriate. Considerable attention has been paid to the formulation of precise definitions and statements of theorems. The proofs of most of the theorems are included in detail in this book.

Many illustrative examples have been included to facilitate the use of the book both for individual study and for summer institutes for teachers of mathematics. This book contains enough material for a one-semester course at the college level or for enrichment programs at the high school level. Also, it may be used together with my book, *Vectors and Their Applications*, for a course in linear algebra. There are a sufficient number of exercises, which range from routine computations to proofs of theorems that extend the theory of the subject. Answers are provided to the odd-numbered exercises.

It is assumed that the reader has some understanding of the basic fundamentals of vector algebra. All of Chapters 1 and 2 may be studied, however, without any previous knowledge of vectors. Certain derivations and interpretations of material contained in Chapters 3 and 4 may be omitted by the reader who is unfamiliar with geometric vectors.

To the many students and teachers who have contributed to my understanding of matrix theory and to the preparation of this book, I acknowledge a debt of gratitude. I am particularly grateful to Dr. Bruce E. Meserve of the University of Vermont for suggesting that this book be written and for his invaluable comments and criticisms throughout its development. I wish to express my appreciation to my wife, Betty, for typing the final manuscript as well as the preliminary versions. A special note of appreciation is due the editorial-production staff of Prentice-Hall, Inc., for their cooperation in the production of this book.

Anthony J. Pettofrezzo

Contents

1 MATRICES   1

1–1 Definitions and Elementary Properties   1
1–2 Matrix Multiplication   6
1–3 Diagonal Matrices   13
1–4 Special Real Matrices   15
1–5 Special Complex Matrices   19

2 INVERSES AND SYSTEMS OF MATRICES   22

2–1 Determinants   22
2–2 Inverse of a Matrix   28
2–3 Systems of Matrices   35
2–4 Rank of a Matrix   41
2–5 Systems of Linear Equations   46

3 TRANSFORMATIONS OF THE PLANE   51

3–1 Mappings   51
3–2 Rotations   53
3–3 Reflections, Dilations, and Magnifications   58
3–4 Other Transformations   63
3–5 Linear Homogeneous Transformations   66

3–6 Orthogonal Matrices   68
3–7 Translations   71
3–8 Rigid Motion Transformations   76

4 Eigenvalues and Eigenvectors   83

4–1 Characteristic Functions   83
4–2 A Geometric Interpretation of Eigenvectors   87
4–3 Some Theorems   89
4–4 Diagonalization of Matrices   92
4–5 The Hamilton-Cayley Theorem   97
4–6 Quadratic Forms   101
4–7 Classification of the Conics   103
4–8 Invariants for Conics   109

Bibliography   112

Answers to Odd-Numbered
Exercises   114

Index   129

chapter 1

Matrices

1-1 Definitions and Elementary Properties

In many branches of the physical, biological, and social sciences it is necessary for scientists to express and use a set of numbers in a rectangular array. Indeed, in many everyday activities it is convenient, if not necessary, to use sets of numbers arranged in rows and columns for keeping records, for purposes of comparison, and for a variety of other reasons.

Consider a company that manufactures three models of typewriters: an electric model, a standard model, and a portable model. If the company wishes to compare the units of raw material and labor involved in one month's production of each of these models, an *array* may be used to present the data:

$$\begin{array}{c} & \begin{array}{ccc} \textit{Electric} & \textit{Standard} & \textit{Portable} \\ \textit{model} & \textit{model} & \textit{model} \end{array} \\ \begin{array}{c} \textit{Units of material} \\ \textit{Units of labor} \end{array} & \left(\begin{array}{ccc} 20 & 17 & 12 \\ 6 & 8 & 5 \end{array}\right). \end{array}$$

The units used are not intended to be realistic but merely to illustrate an oversimplified application of an array of real numbers. Units of material for the three models comprise the first row of the array, units of labor the second row, and units of production for each model (material and labor) the columns of the array. If the pattern in which the units are to be recorded is clearly defined in advance, this rectangular array may be presented simply as:

$$\begin{pmatrix} 20 & 17 & 12 \\ 6 & 8 & 5 \end{pmatrix}. \tag{1-1}$$

A second example of the use of rectangular arrays of real numbers is one that might be used by a basketball coach who wishes to keep a record of the scoring performances of three of his players. Consider the following array:

	Games	Field goals	Free throws
Player A	16	110	62
Player B	14	85	42
Player C	16	73	55

or simply

$$\begin{pmatrix} 16 & 110 & 62 \\ 14 & 85 & 42 \\ 16 & 73 & 55 \end{pmatrix}. \tag{1-2}$$

Rectangular arrays of elements a_{ij} such as

$$\begin{pmatrix} a_{11} & a_{12} & \cdots & a_{1n} \\ a_{21} & a_{22} & \cdots & a_{2n} \\ \cdots & \cdots & \cdots & \cdots \\ a_{m1} & a_{m2} & \cdots & a_{mn} \end{pmatrix} \tag{1-3}$$

are called **matrices** (singular: **matrix**). Each element a_{ij} has two indices: the **row index**, i, and **column index**, j. The elements $a_{i1}, a_{i2}, \ldots, a_{in}$ are the elements of the ith row, and the elements $a_{1j}, a_{2j}, \ldots, a_{mj}$ are the elements of the jth column. The element a_{ij} is the element contained simultaneously in the ith row and jth column. For example, the element a_{21} of matrix (1-1) is equal to 6; that is, a_{21} is the element in the second row and first column.

A matrix of m rows and n columns is called a matrix of **order** m by n. Thus, matrix (1-1) is of order 2 by 3, while matrix (1-2) is of order 3 by 3. In general, when the number of rows equals the number of columns, the matrix is called a **square matrix**. A square matrix of order n by n is said, simply, to be of order n. Matrix (1-2) is an example of a square matrix of order three.

If each of the elements of a matrix is a real number, the matrix is called a **real matrix**. Unless otherwise stated, we shall be concerned only with real matrices.

Whenever it is convenient, matrices will be denoted symbolically by capital letters $A, B, C, \ldots$, or by $((a_{ij})), ((b_{ij})), ((c_{ij})), \ldots$ where $a_{ij}, b_{ij}, c_{ij}, \ldots$, respectively, represent the general elements of the matrices.

Example 1 Construct a square matrix $((a_{ij}))$ of order three where $a_{ij} = 3i - j^2$.

If $a_{ij} = 3i - j^2$, then

$a_{11} = 3(1) - (1)^2 = 2,\ a_{12} = 3(1) - (2)^2 = -1,\ a_{13} = 3(1) - (3)^2 = -6,$

$a_{21} = 3(2) - (1)^2 = 5,\ a_{22} = 3(2) - (2)^2 = 2,\qquad a_{23} = 3(2) - (3)^2 = -3,$

$a_{31} = 3(3) - (1)^2 = 8,\ a_{32} = 3(3) - (2)^2 = 5,\qquad a_{33} = 3(3) - (3)^2 = 0.$

Hence, the desired matrix is

$$\begin{pmatrix} 2 & -1 & -6 \\ 5 & 2 & -3 \\ 8 & 5 & 0 \end{pmatrix}.$$

Two matrices $((a_{ij}))$ and $((b_{ij}))$ are said to be **equal** if and only if they are of the same order, and $a_{ij} = b_{ij}$ for all pairs (i, j).

Example 2 Determine whether or not the matrices of each pair are equal:

(a) $\begin{pmatrix} 4 & 1 & 2 \\ -2 & 3 & 5 \end{pmatrix}$ and $\begin{pmatrix} 1 & 2 \\ 3 & 5 \end{pmatrix}$;   (b) $\begin{pmatrix} 5 & 2 \\ 1 & 4 \end{pmatrix}$ and $\begin{pmatrix} 5 & 2 \\ 1 & 4 \end{pmatrix}$;

(c) $\begin{pmatrix} a \\ b \\ c \end{pmatrix}$ and $\begin{pmatrix} 2a \\ 2b \\ 2c \end{pmatrix}$;   (d) $\begin{pmatrix} 2 & -2 \\ 1 & 1 \\ 3 & 0 \end{pmatrix}$ and $\begin{pmatrix} 0 & -2 \\ 0 & 1 \\ 0 & 0 \end{pmatrix}.$

The matrices in (a) cannot be equal since they are not of the same order. The matrices in (b) are equal. The matrices in (c) are equal if and only if $a = b = c = 0$. Although the matrices in (d) are of the same order, they are not equal, since not all corresponding elements are equal.

Consider again matrix (1-2) which represents the scoring performances of three basketball players for one season. Suppose that the matrix representing the scoring performances of these players in the next season of play is

$$\begin{pmatrix} 18 & 142 & 98 \\ 18 & 83 & 33 \\ 15 & 103 & 60 \end{pmatrix}.$$

A matrix representing the combined scoring performance of each of the three players during two seasons may be obtained by adding the corresponding entries of the two matrices:

$$\begin{pmatrix} 16 & 110 & 62 \\ 14 & 85 & 42 \\ 16 & 73 & 55 \end{pmatrix} + \begin{pmatrix} 18 & 142 & 98 \\ 18 & 83 & 30 \\ 15 & 103 & 60 \end{pmatrix} = \begin{pmatrix} 34 & 252 & 160 \\ 32 & 168 & 72 \\ 31 & 176 & 115 \end{pmatrix}.$$

That is, in two seasons players A, B, and C participated in 34, 32, and 31 games, scored 252, 168, and 176 field goals, and 160, 72, and 115 free throws, respectively.

In general, the addition of two matrices $((a_{ij}))$ and $(b_{ij}))$ is defined if and only if the matrices are of the same order. If $((a_{ij}))$ and $((b_{ij}))$ are matrices of the same order, then the **sum** $((a_{ij})) + ((b_{ij}))$ is defined as a third matrix $((c_{ij}))$ of that same order where each element c_{ij} satisfies the condition $c_{ij} = a_{ij} + b_{ij}$.

Consider any three real matrices of order m by n: $A = ((a_{ij}))$, $B = ((b_{ij}))$, and $C = ((c_{ij}))$. Since the addition of real numbers is commutative, $a_{ij} + b_{ij} = b_{ij} + a_{ij}$ for all pairs (i, j) and

$$A + B = B + A. \tag{1-4}$$

Since the addition of real numbers is associative, $(a_{ij} + b_{ij}) + c_{ij} = a_{ij} + (b_{ij} + c_{ij})$ for all pairs (i, j) and

$$(A + B) + C = A + (B + C). \tag{1-5}$$

Thus, *the addition of real matrices is commutative and associative.*

A **null matrix** or **zero matrix,** denoted by 0, is a matrix wherein all of the elements are zero. For every matrix A of order m by n there exists a zero matrix of order m by n such that $A + 0 = 0 + A = A$. This zero matrix of order m by n is the *additive identity element* for the set of all matrices of order m by n.

Example 3 Find the sum of matrices A and B where

$$A = \begin{pmatrix} 2 & 1 & 1 \\ 0 & -3 & 4 \end{pmatrix} \quad \text{and} \quad B = \begin{pmatrix} 3 & -1 & 3 \\ 2 & 0 & 5 \end{pmatrix}.$$

$$A + B = \begin{pmatrix} 2+3 & 1+(-1) & 1+3 \\ 0+2 & -3+0 & 4+5 \end{pmatrix} = \begin{pmatrix} 5 & 0 & 4 \\ 2 & -3 & 9 \end{pmatrix}.$$

Consider again matrix (1-1) which represents the units of material and labor involved in one month's production of three models of typewriters. Suppose the manufacturing company wishes to double its production of each model. Then the matrix

$$\begin{pmatrix} 40 & 34 & 24 \\ 12 & 16 & 10 \end{pmatrix}$$

would represent the units of material and labor involved in a single month's production of the three models of typewriters. It is convenient to represent the doubling of the entries in matrix (1-1) as a product of the matrix and the real number 2; that is,

$$2\begin{pmatrix} 20 & 17 & 12 \\ 6 & 8 & 5 \end{pmatrix} = \begin{pmatrix} 40 & 34 & 24 \\ 12 & 16 & 10 \end{pmatrix}.$$

Note that

$$2\begin{pmatrix} 20 & 17 & 12 \\ 6 & 8 & 5 \end{pmatrix} = \begin{pmatrix} 20 & 17 & 12 \\ 6 & 8 & 5 \end{pmatrix} + \begin{pmatrix} 20 & 17 & 12 \\ 6 & 8 & 5 \end{pmatrix}.$$

In general, the product of a real number (scalar) k and a matrix $((a_{ij}))$, denoted by $k((a_{ij}))$ or by $((a_{ij}))k$, is called the scalar multiple of the matrix $((a_{ij}))$ by k. The **scalar multiple** $k((a_{ij}))$ is defined as a matrix wherein the elements are products of k and the corresponding elements of $((a_{ij}))$. Since the multiplication of real numbers is commutative,

$$k((a_{ij})) = ((a_{ij}))k = ((ka_{ij})). \tag{1-6}$$

Notice that for any real number k and any real matrix $A = ((a_{ij}))$, the matrices A and kA are of the same order. In particular, if $k = -1$, then

$$A + (-1)A = ((a_{ij})) + ((-a_{ij})) = 0.$$

Thus, $(-1)A$ is the *additive inverse* of A. If B and A are any two matrices of the same order, the **difference** $B - A$ is defined by the relation

$$B - A = B + (-1)A. \tag{1-7}$$

In general, the scalars may be considered as scalar coefficients, and any algebraic sum of scalar multiples of matrices of the same order satisfies certain laws. For example, if k and l are scalars and A and B are matrices of the same order, then

$$kA + lA = (k + l)A, \tag{1-8}$$

$$klA = k(lA) = l(kA) = (kl)A, \tag{1-9}$$

and

$$k(A + B) = kA + kB. \tag{1-10}$$

Furthermore, if $kA = 0$, then either $k = 0$ or A is a zero matrix.

Example 4 Find $3A - 2B$ where

$$A = \begin{pmatrix} 1 & 2 \\ 3 & 0 \end{pmatrix} \quad \text{and} \quad B = \begin{pmatrix} 1 & 3 \\ 0 & -4 \end{pmatrix}.$$

$$3A - 2B = 3\begin{pmatrix} 1 & 2 \\ 3 & 0 \end{pmatrix} - 2\begin{pmatrix} 1 & 3 \\ 0 & -4 \end{pmatrix} = 3\begin{pmatrix} 1 & 2 \\ 3 & 0 \end{pmatrix} + (-2)\begin{pmatrix} 1 & 3 \\ 0 & -4 \end{pmatrix}$$

$$= \begin{pmatrix} 3(1) & 3(2) \\ 3(3) & 3(0) \end{pmatrix} + \begin{pmatrix} -2(1) & -2(3) \\ -2(0) & -2(-4) \end{pmatrix}$$

$$= \begin{pmatrix} 3 & 6 \\ 9 & 0 \end{pmatrix} + \begin{pmatrix} -2 & -6 \\ 0 & 8 \end{pmatrix} = \begin{pmatrix} 1 & 0 \\ 9 & 8 \end{pmatrix}.$$

Exercises

1. Construct a square matrix $((a_{ij}))$ of order three where $a_{ij} = i^2 + 2j - 3$.

2. Construct a matrix $((a_{ij}))$ of order 3 by 2 where $a_{ij} = i^2 - ij$.

3. In the square matrix $((a_{ij}))$ of order two describe the position of the elements for which (a) $i = 2$, (b) $j = 1$, and (c) $i = j$.

4. If

$$A = \begin{pmatrix} 2 & -1 & 5 \\ 3 & 2 & 1 \end{pmatrix} \quad \text{and} \quad B = \begin{pmatrix} -2 & 0 & 3 \\ 1 & -4 & -1 \end{pmatrix},$$

then find **(a)** $A + B$; **(b)** $A - B$; **(c)** $A + 3B$.

5. Verify the associative law of addition of matrices (1-5) for

$$A = \begin{pmatrix} 3 & 1 & 1 \\ -2 & 5 & 0 \end{pmatrix}, \quad B = \begin{pmatrix} 1 & -3 & 0 \\ 0 & 1 & 4 \end{pmatrix}, \quad \text{and} \quad C = \begin{pmatrix} 2 & 7 & -1 \\ -2 & 1 & 3 \end{pmatrix}.$$

6. Find the additive inverse of the matrix

(a) $\begin{pmatrix} 2 & 5 & 3 \\ 0 & 2 & -1 \end{pmatrix};$ **(b)** $\begin{pmatrix} a & b \\ c & d \end{pmatrix}.$

7. Solve the matrix equation

$$\begin{pmatrix} a_{11} & a_{12} \\ a_{21} & a_{22} \end{pmatrix} + \begin{pmatrix} 3 & 1 \\ -4 & 0 \end{pmatrix} = \begin{pmatrix} -3 & 2 \\ 1 & 0 \end{pmatrix}.$$

8. Prove that the set of real matrices of order m by n forms a commutative group under matrix addition.

1-2 Matrix Multiplication

Consider a system of linear equations such as

$$\begin{cases} 2x - y + 2z = 1 \\ x + 2y - 4z = 3 \\ 3x - y + z = 0. \end{cases} \tag{1-11}$$

This system may be represented by a single matrix equation:

$$\begin{pmatrix} 2x - y + 2z \\ x + 2y - 4z \\ 3x - y + z \end{pmatrix} = \begin{pmatrix} 1 \\ 3 \\ 0 \end{pmatrix}. \tag{1-12}$$

The coefficients of x, y, and z may be obtained either from (1-11) or (1-12). In both cases the solution depends upon these coefficients. The **matrix of coefficients** is

$$\begin{pmatrix} 2 & -1 & 2 \\ 1 & 2 & -4 \\ 3 & -1 & 1 \end{pmatrix}.$$

The coefficients of each variable are positioned in a column, and coefficients of the variables of each equation are located on a row. It is customary and convenient to think of this matrix of coefficients as an operator that acts upon a column matrix of the variables:

$$\begin{pmatrix} 2 & -1 & 2 \\ 1 & 2 & -4 \\ 3 & -1 & 1 \end{pmatrix} \begin{pmatrix} x \\ y \\ z \end{pmatrix} = \begin{pmatrix} 2x - y + 2z \\ x + 2y - 4z \\ 3x - y + z \end{pmatrix}. \tag{1-13}$$

Then, the system of equations (1-11) may be represented by the single matrix equation

$$\begin{pmatrix} 2 & -1 & 2 \\ 1 & 2 & -4 \\ 3 & -1 & 1 \end{pmatrix} \begin{pmatrix} x \\ y \\ z \end{pmatrix} = \begin{pmatrix} 1 \\ 3 \\ 0 \end{pmatrix}. \tag{1-14}$$

Use of the matrix of coefficients as an operator in (1-13) requires the introduction of *matrix multiplication*. Notice that the element $2x - y + 2z$ may be obtained from the matrices

$$\begin{pmatrix} 2 & -1 & 2 \\ 1 & 2 & -4 \\ 3 & -1 & 1 \end{pmatrix} \quad \text{and} \quad \begin{pmatrix} x \\ y \\ z \end{pmatrix}$$

by summing the products of the elements of row one of the matrix of coefficients and the corresponding elements of the column matrix of variables, taken in order; that is,

$$(2)(x) + (-1)(y) + (2)(z) = 2x - y + 2z.$$

Similarly, the element $x + 2y - 4z$ may be obtained by summing the products of the elements of row two of the matrix of coefficients and the corresponding elements of the column matrix of variables, taken in order; that is,

$$(1)(x) + (2)(y) + (-4)(z) = x + 2y - 4z.$$

In a similar manner, using the elements of row three of the matrix of coefficients, we obtain

$$(3)(x) + (-1)(y) + (1)(z) = 3x - y + z.$$

In general, the **product** AB of two matrices A and B is defined to be a matrix C such that the element in the ith row and jth column of C is obtained by summing the products of the elements of the ith row of A and the corresponding elements of the jth column of B, taken in order. Notice that the number of columns of A must be the same as the number of rows of B. If $A = ((a_{ij}))$ is a matrix of order m by n and $B = ((b_{ij}))$ is a matrix of order n by r, then $C = ((c_{ij}))$ is a matrix of order m by r where

$$c_{ij} = a_{i1}b_{1j} + a_{i2}b_{2j} + \cdots + a_{in}b_{nj}. \tag{1-15}$$

We may use summation notation and write (1-15) as

$$c_{ij} = \sum_{k=1}^{n} a_{ik}b_{kj}. \tag{1-16}$$

When the number of columns of a matrix A is equal to the number of rows of a matrix B, the product AB exists, and the matrices A and B are said to be **conformable** for the product AB. Two matrices can be multiplied only when they are conformable. In the product AB, B is sometimes spoken of as being **premultiplied** by A, and A as being **postmultiplied** by B. Even if the product AB exists, the product BA may not exist since matrices A and B may not be conformable for product BA. This illustrates an important property of matrix multiplication, namely, that it is, in general, not commutative.

Example 1 Find the products AB and BA, if they exist, where

$$A = \begin{pmatrix} 2 & 3 \\ 1 & -4 \end{pmatrix} \quad \text{and} \quad B = \begin{pmatrix} 3 & -2 & 2 \\ 1 & 0 & -1 \end{pmatrix}.$$

Matrices A and B are conformable for the product AB since the number of columns of A is equal to the number of rows of B. Hence, AB exists. Furthermore, AB is of order 2 by 3 since A is of order 2 by 2 and B is of order 2 by 3. By definition, the general element of AB is given as $c_{ij} = a_{i1}b_{1j} + a_{i2}b_{2j}$. Then

$$c_{11}=(2)(3)+(3)(1), \quad c_{12}=(2)(-2)+(3)(0), \quad c_{13}=(2)(2)+(3)(-1),$$
$$c_{21}=(1)(3)+(-4)(1), \quad c_{22}=(1)(-2)+(-4)(0), \quad c_{23}=(1)(2)+(-4)(-1);$$

that is,

$$c_{11} = 9, \quad c_{12} = -4, \quad c_{13} = 1,$$
$$c_{21} = -1, \quad c_{22} = -2, \quad c_{23} = 6.$$

Therefore,

$$AB = \begin{pmatrix} 9 & -4 & 1 \\ -1 & -2 & 6 \end{pmatrix}.$$

The product BA does not exist since the matrices B and A are not conformable for the product BA.

If the elements of any row or column of a matrix are considered to represent the components of a vector; for every pair of values (i, j), the element c_{ij} of (1-15) is the scalar (sometimes called the dot or inner) product of the ith row vector of A and the jth column vector of B. A matrix consisting of a single row is sometimes called a **row matrix** or a **row vector**; a matrix consisting of a single column is sometimes called a **column matrix** or a **column vector.**

Example 2 Find the matrix products AB and BA of the row vector $A = (1 \quad 2 \quad 3)$ and the column vector

$$B = \begin{pmatrix} -2 \\ 4 \\ 1 \end{pmatrix}.$$

Since A is of order 1 by 3 and B is of order 3 by 1, the matrices are conformable regardless of the order in which they are considered. Hence, the products AB and BA both exist:

$$AB = ((1)(-2) + (2)(4) + (3)(1)) = (9);$$

$$BA = \begin{pmatrix} (-2)(1) & (-2)(2) & (-2)(3) \\ (4)(1) & (4)(2) & (4)(3) \\ (1)(1) & (1)(2) & (1)(3) \end{pmatrix} = \begin{pmatrix} -2 & -4 & -6 \\ 4 & 8 & 12 \\ 1 & 2 & 3 \end{pmatrix}.$$

Note that the product AB may be considered a matrix whose only element represents the scalar product of two vectors whose components are the elements of A and B, respectively.

Example 3 Prove that $C(A + B) = CA + CB$ where

$$A = \begin{pmatrix} 1 & 2 \\ 3 & 0 \end{pmatrix}, \quad B = \begin{pmatrix} 2 & -1 \\ 3 & 4 \end{pmatrix}, \quad \text{and} \quad C = \begin{pmatrix} 2 & -2 \\ 1 & 3 \\ 4 & -1 \end{pmatrix}.$$

$$C(A + B) = \begin{pmatrix} 2 & -2 \\ 1 & 3 \\ 4 & -1 \end{pmatrix} \left[\begin{pmatrix} 1 & 2 \\ 3 & 0 \end{pmatrix} + \begin{pmatrix} 2 & -1 \\ 3 & 4 \end{pmatrix} \right]$$

$$= \begin{pmatrix} 2 & -2 \\ 1 & 3 \\ 4 & -1 \end{pmatrix} \begin{pmatrix} 3 & 1 \\ 6 & 4 \end{pmatrix} = \begin{pmatrix} -6 & -6 \\ 21 & 13 \\ 6 & 0 \end{pmatrix};$$

$$CA + CB = \begin{pmatrix} 2 & -2 \\ 1 & 3 \\ 4 & -1 \end{pmatrix} \begin{pmatrix} 1 & 2 \\ 3 & 0 \end{pmatrix} + \begin{pmatrix} 2 & -2 \\ 1 & 3 \\ 4 & -1 \end{pmatrix} \begin{pmatrix} 2 & -1 \\ 3 & 4 \end{pmatrix}$$

$$= \begin{pmatrix} -4 & 4 \\ 10 & 2 \\ 1 & 8 \end{pmatrix} + \begin{pmatrix} -2 & -10 \\ 11 & 11 \\ 5 & -8 \end{pmatrix} = \begin{pmatrix} -6 & -6 \\ 21 & 13 \\ 6 & 0 \end{pmatrix};$$

hence, $C(A + B) = CA + CB$.

Example 3 is an illustration of a general theorem of matrix algebra.

Theorem 1-1 *The multiplication of matrices is distributive with respect to addition.*

Proof: Let $A = ((a_{ij}))$ and $B = ((b_{ij}))$ be matrices of order m by n, and let $C = ((c_{ij}))$ be a matrix of order k by m. Then $A + B$, CA, CB, and $C(A + B)$ exist. The elements of the ith row of C are

$$c_{i1}, c_{i2}, \ldots, c_{im},$$

and the elements of the jth column of $A + B$ are

$$a_{1j} + b_{1j}, a_{2j} + b_{2j}, \ldots, a_{mj} + b_{mj}.$$

Therefore, the ijth element of $C(A + B)$ is

$$c_{i1}(a_{1j} + b_{1j}) + c_{i2}(a_{2j} + b_{2j}) + \cdots + c_{im}(a_{mj} + b_{mj});$$

that is,

$$(c_{i1}a_{1j} + c_{i2}a_{2j} + \cdots + c_{im}a_{mj}) + (c_{i1}b_{1j} + c_{i2}b_{2j} + \cdots + c_{im}b_{mj}),$$

the sum of the ijth elements of CA and CB, respectively. Hence,

$$C(A + B) = CA + CB. \qquad (1\text{-}17)$$

Equation (1-17) represents the *left-hand distributive property* of matrix multiplication. The *right-hand distributive property*

$$(A + B)C = AC + BC \qquad (1\text{-}18)$$

also is valid provided $A + B$, AC, BC, and $(A + B)C$ exist. Note that $C(A + B)$ and $(A + B)C$ generally are not equal.

Example 4 Prove that $A(BC) = (AB)C$ where

$$A = \begin{pmatrix} 1 & 2 \\ -1 & 3 \end{pmatrix}, \quad B = \begin{pmatrix} 1 & 0 & -1 \\ 2 & 1 & 0 \end{pmatrix}, \quad \text{and} \quad C = \begin{pmatrix} 1 & -1 \\ 3 & 2 \\ 2 & 1 \end{pmatrix}.$$

$$A(BC) = \begin{pmatrix} 1 & 2 \\ -1 & 3 \end{pmatrix} \left[\begin{pmatrix} 1 & 0 & -1 \\ 2 & 1 & 0 \end{pmatrix} \begin{pmatrix} 1 & -1 \\ 3 & 2 \\ 2 & 1 \end{pmatrix} \right]$$

$$= \begin{pmatrix} 1 & 2 \\ -1 & 3 \end{pmatrix} \begin{pmatrix} -1 & -2 \\ 5 & 0 \end{pmatrix} = \begin{pmatrix} 9 & -2 \\ 16 & 2 \end{pmatrix};$$

$$(AB)C = \left[\begin{pmatrix} 1 & 2 \\ -1 & 3 \end{pmatrix} \begin{pmatrix} 1 & 0 & -1 \\ 2 & 1 & 0 \end{pmatrix} \right] \begin{pmatrix} 1 & -1 \\ 3 & 2 \\ 2 & 1 \end{pmatrix}$$

$$= \begin{pmatrix} 5 & 2 & -1 \\ 5 & 3 & 1 \end{pmatrix} \begin{pmatrix} 1 & -1 \\ 3 & 2 \\ 2 & 1 \end{pmatrix} = \begin{pmatrix} 9 & -2 \\ 16 & 2 \end{pmatrix};$$

hence, $A(BC) = (AB)C$.

Example 4 is an illustration of one of the most important theorems of matrix algebra.

Theorem 1-2 *The multiplication of matrices is associative.*

Proof: Let $A = ((a_{ij}))$, $B = ((b_{ij}))$, and $C = ((c_{ij}))$ be matrices of order k by m, m by n, and n by p, respectively. Then the products AB, BC, $A(BC)$, and $(AB)C$ exist. The elements of the ith row of A are a_{i1}, $a_{i2}, \ldots, a_{im}$, and the elements of the jth column of BC are $b_{11}c_{1j} + b_{12}c_{2j}$

$$+ \cdots + b_{1n}c_{nj}, \; b_{21}c_{1j} + b_{22}c_{2j} + \cdots + b_{2n}c_{nj}, \ldots, \; b_{m1}c_{1j} + b_{m2}c_{2j}$$
$+ \cdots + b_{mn}c_{nj}$. Therefore, the ijth element of $A(BC)$ is

$$a_{i1}(b_{11}c_{1j} + b_{12}c_{2j} + \cdots + b_{1n}c_{nj}) +$$
$$a_{i2}(b_{21}c_{1j} + b_{22}c_{2j} + \cdots + b_{2n}c_{nj}) +$$
$$\cdots + a_{im}(b_{m1}c_{1j} + b_{m2}c_{2j} + \cdots + b_{mn}c_{nj});$$

that is,

$$(a_{i1}b_{11} + a_{i2}b_{21} + \cdots + a_{im}b_{m1})c_{1j} +$$
$$(a_{i1}b_{12} + a_{i2}b_{22} + \cdots + a_{im}b_{m2})c_{2j} +$$
$$\cdots + (a_{i1}b_{1n} + a_{i2}b_{2n} + \cdots + a_{im}b_{mn})c_{nj},$$

which represents the ijth element of $(AB)C$. Hence,

$$A(BC) = (AB)C. \tag{1-19}$$

Consider the matrices

$$A = \begin{pmatrix} 1 & 0 \\ 1 & 0 \end{pmatrix} \quad \text{and} \quad B = \begin{pmatrix} 0 & 0 \\ 1 & 1 \end{pmatrix}.$$

Note that $AB = 0$, but neither A nor B is a zero matrix. In matrix algebra it is possible to have **zero divisors.** These are nonzero elements whose product is zero. In the algebra of real numbers it can be shown that zero divisors do not exist; that is, if a and b are real numbers and $ab = 0$, then $a = 0$ or $b = 0$.

Exercises

1. Find AB and BA, if they exist, where

$$A = \begin{pmatrix} 3 & 4 & 0 \\ -1 & 0 & 2 \end{pmatrix} \quad \text{and} \quad B = \begin{pmatrix} 6 & -1 & 2 \\ 0 & 1 & 5 \\ -1 & 3 & 4 \end{pmatrix}.$$

2. Use the matrices given to verify that the multiplication of square matrices generally is not commutative:

$$A = \begin{pmatrix} -1 & 2 \\ 3 & 1 \end{pmatrix} \quad \text{and} \quad B = \begin{pmatrix} 0 & 2 \\ 1 & 4 \end{pmatrix}.$$

3. Verify the associative property of matrix multiplication (1-19) for

$$A = \begin{pmatrix} 2 & 1 \\ -1 & 0 \end{pmatrix}, \quad B = \begin{pmatrix} 3 & 1 \\ 2 & -2 \end{pmatrix}, \quad \text{and} \quad C = \begin{pmatrix} 0 & 1 \\ 2 & 1 \end{pmatrix}.$$

4. Verify the right-hand distributive property (1-18) for

$$A = \begin{pmatrix} 1 & 2 & 3 \\ -1 & 0 & 4 \end{pmatrix}, \quad B = \begin{pmatrix} 2 & 0 & 2 \\ 1 & 3 & -1 \end{pmatrix}, \quad \text{and} \quad C = \begin{pmatrix} 2 & 1 \\ -1 & 0 \\ 3 & 2 \end{pmatrix}.$$

5. Find:

$$\begin{pmatrix} \cos\theta & \sin\theta \\ -\sin\theta & \cos\theta \end{pmatrix} \begin{pmatrix} \cos\theta & -\sin\theta \\ \sin\theta & \cos\theta \end{pmatrix}.$$

6. Determine conditions on m, n, and p such that the product AB of matrix A of order m by n and matrix B of order n by p is a square matrix.

7. Given

$$A = \begin{pmatrix} 1 & 3 \\ -2 & 0 \end{pmatrix} \quad \text{and} \quad B = \begin{pmatrix} 2 & 1 \\ -1 & 2 \end{pmatrix},$$

show that $(A + B)(A - B) \neq A^2 - B^2$.

8. Determine a necessary and sufficient condition for
    (a) $(A + B)(A + B) = A^2 + B^2$;   **(b)** $(A + B)(A - B) = A^2 - B^2$.

9. Find all matrices A such that

  (a) $\begin{pmatrix} 3 & 0 \\ 5 & 0 \end{pmatrix} A = \begin{pmatrix} 2 & 0 & 0 \\ 1 & 0 & 0 \end{pmatrix};$ **(b)** $\begin{pmatrix} 2 & 1 \\ 3 & 2 \end{pmatrix} A = A \begin{pmatrix} 2 & 1 \\ 3 & 2 \end{pmatrix}.$

10. Prove that $AB = BA$ where

$$A = \begin{pmatrix} r & s \\ -s & r \end{pmatrix} \quad \text{and} \quad B = \begin{pmatrix} m & n \\ -n & m \end{pmatrix}.$$

11. Find A^3 if $A^3 = A\,(AA)$ and

$$A = \begin{pmatrix} 2 & 3 \\ 1 & 0 \end{pmatrix}.$$

12. If

$$A = \begin{pmatrix} 1 & 1 \\ 1 & 0 \end{pmatrix},$$

discuss the nature of A^n where n is a positive integer. (Note: The sequence 1, 1, 2, 3, 5, 8, 13, 21, 34, . . . , where each term is the sum of the two preceding terms, is called a *Fibonacci sequence*.)

13. If $AC = CA$ and $BC = CB$, prove that $C(AB + BA) = (AB + BA)C$.

14. Prove that the set of *Pauli matrices*,

$$A = \begin{pmatrix} 1 & 0 \\ 0 & 1 \end{pmatrix}, \quad B = \begin{pmatrix} 0 & 1 \\ -1 & 0 \end{pmatrix}, \quad C = \begin{pmatrix} 0 & -1 \\ 1 & 0 \end{pmatrix}, \quad D = \begin{pmatrix} -1 & 0 \\ 0 & -1 \end{pmatrix},$$

$$E = \begin{pmatrix} i & 0 \\ 0 & -i \end{pmatrix}, \quad F = \begin{pmatrix} -i & 0 \\ 0 & i \end{pmatrix}, \quad G = \begin{pmatrix} 0 & -i \\ -i & 0 \end{pmatrix}, \quad H = \begin{pmatrix} 0 & i \\ i & 0 \end{pmatrix},$$

forms a group under matrix multiplication. The Pauli matrices are used in the study of atomic physics.

1-3 Diagonal Matrices

The elements a_{ij} where $i = j$ of a square matrix $((a_{ij}))$ are called the **diagonal elements** of $((a_{ij}))$ and are said to be on the **main diagonal** or **principal diagonal.** A square matrix of the form

$$A = \begin{pmatrix} a_{11} & 0 & \cdots & 0 \\ 0 & a_{22} & \cdots & 0 \\ \cdots & \cdots & \cdots & \cdots \\ 0 & 0 & \cdots & a_{nn} \end{pmatrix} \tag{1-20}$$

is called a **diagonal matrix**; that is, a diagonal matrix is a square matrix $((a_{ij}))$ where $a_{ij} = 0$ if $i \neq j$ for all pairs (i, j). For example,

$$\begin{pmatrix} 3 & 0 \\ 0 & -1 \end{pmatrix}, \quad \begin{pmatrix} 2 & 0 & 0 \\ 0 & 5 & 0 \\ 0 & 0 & 0 \end{pmatrix}, \quad \text{and} \quad \begin{pmatrix} 3 & 0 & 0 \\ 0 & 3 & 0 \\ 0 & 0 & 3 \end{pmatrix} \tag{1-21}$$

are diagonal matrices as is any zero matrix of order n. If all the a_{ii}'s of (1-20) are equal, then the diagonal matrix is called a **scalar matrix.** The third matrix of (1-21) is an example of a scalar matrix.

Example 1 Determine the effect of the premultiplication and the postmultiplication of any square matrix of order two by **(a)** a conformable diagonal matrix; **(b)** a conformable scalar matrix.

Let

$$A = \begin{pmatrix} a & b \\ c & d \end{pmatrix}$$

be any square matrix of order two,

$$D = \begin{pmatrix} k_1 & 0 \\ 0 & k_2 \end{pmatrix}$$

be a conformable diagonal matrix, and

$$S = \begin{pmatrix} k & 0 \\ 0 & k \end{pmatrix}$$

be a conformable scalar matrix.

(a) By the definition of matrix multiplication,

$$DA = \begin{pmatrix} k_1 & 0 \\ 0 & k_2 \end{pmatrix} \begin{pmatrix} a & b \\ c & d \end{pmatrix} = \begin{pmatrix} k_1a & k_1b \\ k_2c & k_2d \end{pmatrix};$$

that is, premultiplication of matrix A by D results in a matrix where each element of the ith row equals the product of the corresponding element of A and the diagonal element in the ith row of D. Similarly,

$$AD = \begin{pmatrix} a & b \\ c & d \end{pmatrix}\begin{pmatrix} k_1 & 0 \\ 0 & k_2 \end{pmatrix} = \begin{pmatrix} k_1a & k_2b \\ k_1c & k_2d \end{pmatrix};$$

that is, postmultiplication of matrix A by D results in a matrix where each element of the ith column equals the product of the corresponding element of A and the diagonal element in the ith column of D.

(b)
$$SA = \begin{pmatrix} k & 0 \\ 0 & k \end{pmatrix}\begin{pmatrix} a & b \\ c & d \end{pmatrix} = \begin{pmatrix} ka & kb \\ kc & kd \end{pmatrix};$$

$$AS = \begin{pmatrix} a & b \\ c & d \end{pmatrix}\begin{pmatrix} k & 0 \\ 0 & k \end{pmatrix} = \begin{pmatrix} ka & kb \\ kc & kd \end{pmatrix};$$

$SA = AS = kA$. In general, the product of any matrix A and a conformable scalar matrix S with diagonal elements k is equivalent to the scalar multiple kA.

A scalar matrix $((a_{ii}))$ where $a_{ii} = 1$ for all values of i is called an **identity matrix** or a **unit matrix**. An identity matrix of order n will be denoted by I and has the property that for every square matrix A of order n

$$AI = IA = A. \tag{1-22}$$

Assume that I is the only matrix with this property (1-22). Note that if A is a matrix of order m by n, the premultiplicative identity matrix is of order m by m while the postmultiplicative identity matrix is of order n by n.

The identity matrix is often denoted by $((\delta_{ij}))$ where the Kronecker delta symbol δ_{ij} is defined by

$$\delta_{ij} \quad \begin{cases} = 0, & \text{when} \quad i \neq j \\ = 1, & \text{when} \quad i = j. \end{cases} \tag{1-23}$$

Example 2 Show that $AB = BA = I$ where

$$A = \begin{pmatrix} 1 & -1 & 1 \\ 0 & 1 & 0 \\ 2 & 0 & 3 \end{pmatrix} \quad \text{and} \quad B = \begin{pmatrix} 3 & 3 & -1 \\ 0 & 1 & 0 \\ -2 & -2 & 1 \end{pmatrix}.$$

$$AB = \begin{pmatrix} 3+0-2 & 3-1-2 & -1+0+1 \\ 0+0+0 & 0+1+0 & 0+0+0 \\ 6+0-6 & 6+0-6 & -2+0+3 \end{pmatrix} = \begin{pmatrix} 1 & 0 & 0 \\ 0 & 1 & 0 \\ 0 & 0 & 1 \end{pmatrix};$$

$$BA = \begin{pmatrix} 3+0-2 & -3+3+0 & 3+0-3 \\ 0+0+0 & 0+1+0 & 0+0+0 \\ -2+0+2 & 2-2+0 & -2+0+3 \end{pmatrix} = \begin{pmatrix} 1 & 0 & 0 \\ 0 & 1 & 0 \\ 0 & 0 & 1 \end{pmatrix};$$

hence, $AB = BA = I$.

Exercises

1. Construct a matrix $((a_{ij}))$ of order 3 by 4 where $a_{ij} = 3i + \delta_{ij}j^2$.

2. Show that $AB = BA = I$ where

$$A = \begin{pmatrix} 3 & -4 & 2 \\ -2 & 1 & 0 \\ -1 & -1 & 1 \end{pmatrix} \quad \text{and} \quad B = \begin{pmatrix} 1 & 2 & -2 \\ 2 & 5 & -4 \\ 3 & 7 & -5 \end{pmatrix}.$$

3. Show that $A^2 = I$ where

$$A = \begin{pmatrix} -1 & -1 & -1 \\ 0 & 1 & 0 \\ 0 & 0 & 1 \end{pmatrix}.$$

4. Prove that the multiplication of any two diagonal matrices of the same order is commutative.

1-4 Special Real Matrices

A special type of matrix, which shall be used extensively in Chapter 4, is a symmetric matrix. Symmetric matrices play an important role in many branches of mathematics as well as in other sciences. A matrix $((a_{ij}))$ is called a **symmetric matrix** if and only if $a_{ij} = a_{ji}$ for all pairs (i, j). The matrix

$$\begin{pmatrix} 3 & 1 & 0 \\ 1 & 2 & -2 \\ 0 & -2 & 4 \end{pmatrix}$$

is an example of a symmetric matrix of order three.

A matrix $((a_{ij}))$ is called a **skew-symmetric matrix** or **anti-symmetric matrix** if and only if $a_{ij} = -a_{ji}$ for all pairs (i, j). The matrix

$$\begin{pmatrix} 0 & -2 \\ 2 & 0 \end{pmatrix}$$

is an example of a skew-symmetric matrix of order two.

A matrix must necessarily be a square matrix to be symmetric or skew-symmetric. Furthermore, the diagonal elements of a skew-symmetric matrix must be zero since $a_{ii} = -a_{ii}$ if and only if $a_{ii} = 0$.

Example 1 Determine which of the following matrices are symmetric matrices and which are skew-symmetric matrices:

$$A = \begin{pmatrix} 3 & 0 \\ 0 & 2 \end{pmatrix}, \quad B = \begin{pmatrix} 3 & 4 \\ -4 & 1 \end{pmatrix}, \quad C = \begin{pmatrix} 2 & -1 \\ -1 & 1 \end{pmatrix},$$

$$D = \begin{pmatrix} 0 & 2 \\ -2 & 0 \end{pmatrix}, \qquad E = \begin{pmatrix} 0 & 0 \\ 1 & 0 \end{pmatrix}, \qquad F = (3),$$

$$G = \begin{pmatrix} 0 & 1 & -2 \\ -1 & 0 & 3 \\ 2 & -3 & 0 \end{pmatrix}, \quad H = \begin{pmatrix} 3 & 2 \\ 2 & 1 \\ 1 & 0 \end{pmatrix}, \quad J = \begin{pmatrix} 0 & 0 \\ 0 & 0 \end{pmatrix}.$$

Matrices A, C, F, and J are symmetric matrices, while D, G, and J are skew-symmetric matrices. Matrices B, E, and H are neither symmetric nor skew-symmetric. Note that the zero matrix J of order two may be considered both symmetric and skew-symmetric. Zero matrices of any order n are the only matrices that have this property.

Symmetric and skew-symmetric matrices may be discussed in terms of **transposition**, the operation of interchanging the rows and columns of a given matrix. The matrix A^T that is obtained by transposition from a matrix A is called the **transpose** of A. If $A = ((a_{ij}))$, then $A^T = ((a_{ji}))$. Note that the transpose of a column vector is a row vector. In general, the transpose of a matrix of order m by n is a matrix of order n by m. If $a_{ij} = a_{ji}$ for all pairs (i, j) for matrix A, then A is a symmetric matrix and $A^T = A$. If $a_{ij} = -a_{ji}$ for all pairs (i, j) for matrix A, then A is a skew-symmetric matrix and $A^T = -A$.

The following theorems are true for matrices in general.

Theorem 1-3 *The transposition operation is reflexive; that is,* $(A^T)^T = A$.

Proof: Let $A = ((a_{ij}))$ and $A^T = ((b_{ij}))$. Then $b_{ij} = a_{ji}$ for all pairs (i, j). The transpose of A^T is $((b_{ji}))$ and thus $((a_{ij}))$. Hence, $(A^T)^T = A$.

Theorem 1-4 *The transpose of the sum (difference) of two matrices is equal to the sum (difference) of their transposes; that is,* $(A + B)^T = A^T + B^T$ *and* $(A - B)^T = A^T - B^T$.

Proof: Let $A = ((a_{ij}))$ and $B = ((b_{ij}))$ be any two matrices of the same order. Then $A + B = ((c_{ij}))$ where $c_{ij} = a_{ij} + b_{ij}$ for all pairs (i, j);

$$A^T = ((a_{ji})),$$

$$B^T = ((b_{ji})),$$

$$(A + B)^T = ((c_{ji}))$$

$$= ((a_{ji} + b_{ji}))$$

$$= A^T + B^T.$$

In a similar manner, it can be shown that $(A - B)^T = A^T - B^T$.

Theorem 1-5 *The transpose of the product of two matrices is equal to the product of their transposes in reverse order; that is, $(AB)^T = B^T A^T$.*

Proof: Let $A = ((a_{ij}))$ and $B = (b_{ij}))$ be matrices of order k by m and m by n, respectively. Let $AB = ((c_{ij}))$. Then $c_{ij} = a_{i1}b_{1j} + a_{i2}b_{2j} + \cdots + a_{im}b_{mj}$ and is the element of the jth row and ith column of $((c_{ij}))^T$; that is, $(AB)^T$.

The elements $b_{1j}, b_{2j}, \ldots, b_{mj}$ of the jth column of B are the elements of the jth row of B^T. The elements $a_{i1}, a_{i2}, \ldots, a_{im}$ of the ith row of A are the elements of the ith column of A^T. The element of the jth row and ith column of $B^T A^T$ is $b_{1j}a_{i1} + b_{2j}a_{i2} + \cdots + b_{mj}a_{im}$; that is, $a_{i1}b_{1j} + a_{i2}b_{2j} + \cdots + a_{im}b_{mj}$. Hence, $(AB)^T = B^T A^T$.

Theorem 1-6 *The product of any matrix and its transpose is a symmetric matrix; that is, $(AA^T)^T = AA^T$.*

Proof:

$$(AA^T)^T = (A^T)^T A^T \quad \text{by Theorem 1-5}$$
$$= AA^T \quad \text{by Theorem 1-3.}$$

Hence, AA^T is a symmetric matrix.

Note that if A is a matrix of order m by n, then AA^T is a symmetric matrix of order m; $A^T A$ is a symmetric matrix of order n.

The following theorems are true for square matrices.

Theorem 1-7 *The sum of any matrix and its transpose is a symmetric matrix; that is, $(A + A^T)^T = A + A^T$.*

Proof:

$$(A + A^T)^T = A^T + (A^T)^T \quad \text{by Theorem 1-4}$$
$$= A^T + A \quad \text{by Theorem 1-3}$$
$$= A + A^T \quad \text{since the addition of matrices}$$
$$\text{is commutative.}$$

Hence, $A + A^T$ is a symmetric matrix.

Theorem 1-8 *The difference of any matrix and its transpose is a skew-symmetric matrix; that is, $(A - A^T)^T = -(A - A^T)$.*

Proof:

$$(A - A^T)^T = A^T - (A^T)^T \quad \text{by Theorem 1-4}$$
$$= A^T - A \quad \text{by Theorem 1-3}$$
$$= -(A - A^T) \quad \text{by the definitions and properties of scalar}$$
$$\text{multiplication, matrix subtraction, and}$$
$$\text{matrix addition.}$$

Hence, $A - A^T$ is a skew-symmetric matrix.

Theorem 1-9 *If A and B are symmetric matrices, then AB is a symmetric matrix if and only if AB = BA.*

> *Proof:* Let A and B be symmetric matrices of the same order such that AB is a symmetric matrix. Then
>
> $$\begin{aligned} AB &= (AB)^T \quad \text{since } AB \text{ is a symmetric matrix} \\ &= B^T A^T \quad \text{by Theorem 1-5} \\ &= BA \quad \text{since } A \text{ and } B \text{ are symmetric matrices.} \end{aligned}$$
>
> Hence, $AB = BA$.
>
> If $AB = BA$, where A and B are symmetric matrices, then
>
> $$\begin{aligned} (AB)^T &= (BA)^T \\ &= A^T B^T \quad \text{by Theorem 1-5} \\ &= AB \quad \text{since } A \text{ and } B \text{ are symmetric matrices.} \end{aligned}$$
>
> Hence, AB is a symmetric matrix.

Example 2 Verify Theorem 1-5 where

$$A = \begin{pmatrix} 2 & 3 \\ 0 & -1 \end{pmatrix} \quad \text{and} \quad B = \begin{pmatrix} 1 & 5 \\ 2 & 4 \end{pmatrix}.$$

$$AB = \begin{pmatrix} 2 & 3 \\ 0 & -1 \end{pmatrix}\begin{pmatrix} 1 & 5 \\ 2 & 4 \end{pmatrix} = \begin{pmatrix} 8 & 22 \\ -2 & -4 \end{pmatrix}; \quad (AB)^T = \begin{pmatrix} 8 & -2 \\ 22 & -4 \end{pmatrix};$$

$$B^T A^T = \begin{pmatrix} 1 & 2 \\ 5 & 4 \end{pmatrix}\begin{pmatrix} 2 & 0 \\ 3 & -1 \end{pmatrix} = \begin{pmatrix} 8 & -2 \\ 22 & -4 \end{pmatrix}; \quad (AB)^T = B^T A^T.$$

Exercises

1. Determine which of the following matrices are
   (a) symmetric matrices;   (b) skew-symmetric matrices:

$$A = \begin{pmatrix} 2 & 3 \\ 4 & 5 \end{pmatrix}, \qquad B = \begin{pmatrix} 3 & 1 \\ -1 & 3 \end{pmatrix}, \qquad C = \begin{pmatrix} 2 & 2 \\ 2 & 2 \end{pmatrix},$$

$$D = \begin{pmatrix} 5 & 1 \\ 1 & 2 \end{pmatrix}, \qquad E = \begin{pmatrix} 5 & 2 & 1 \\ 4 & 2 & 4 \\ 1 & 2 & 3 \end{pmatrix}, \qquad F = \begin{pmatrix} 0 & -3 \\ 3 & 0 \end{pmatrix},$$

$$G = \begin{pmatrix} 1 & 1 & 1 \\ 2 & 2 & 2 \\ 3 & 3 & 3 \end{pmatrix}, \qquad H = \begin{pmatrix} 0 \\ 0 \\ 0 \end{pmatrix}, \qquad J = \begin{pmatrix} 0 & 0 & 0 \\ 0 & 0 & 0 \\ 0 & 0 & 0 \end{pmatrix}.$$

2. Determine the maximum number of distinct elements in any symmetric matrix of order n.

3. Determine the maximum number of distinct elements in any skew-symmetric matrix of order n.

4. Verify Theorem 1-5 for matrices A and B of Exercise 1.

5. Verify Theorem 1-6 for matrix E of Exercise 1.

6. Verify Theorem 1-7 for matrix A of Exercise 1.

7. Verify Theorem 1-8 for matrix E of Exercise 1.

8. Prove that every diagonal matrix is a symmetric matrix.

9. Prove that the square of a skew-symmetric matrix is a symmetric matrix.

10. Verify the results of Exercise 9 for the matrix

$$A = \begin{pmatrix} 0 & a & b \\ -a & 0 & -c \\ -b & c & 0 \end{pmatrix}.$$

11. Prove that if A and B are skew-symmetric matrices of the same order, AB is a symmetric matrix if and only if $AB = BA$.

1-5 Special Complex Matrices

This book is concerned primarily with real matrices; however, some special complex matrices, which will be useful in Chapter 4 for proving a set of theorems about real symmetric matrices, are considered in this section. A **complex matrix** is a matrix whose elements are complex numbers. Since every real number is a complex number, every real matrix is a complex matrix, but not every complex matrix is a real matrix.

If A is a complex matrix, then $\bar{A}$ denotes the matrix obtained from A by replacing each element $z = a + bi$ with its conjugate $\bar{z} = a - bi$. The matrix $\bar{A}$ is called the **conjugate** of matrix A. For example, each of the matrices

$$\begin{pmatrix} 2+i & 3 \\ i & 5-2i \end{pmatrix} \quad \text{and} \quad \begin{pmatrix} 2-i & 3 \\ -i & 5+2i \end{pmatrix}$$

is the conjugate of the other. Note that matrix A is a real matrix if and only if $A = \bar{A}$. The transpose of the conjugate of matrix A will be denoted by A^*; that is, $A^* = (\bar{A})^T$. If $A = ((a_{ij}))$, then $A^T = ((a_{ji}))$, $\bar{A} = ((\overline{a_{ij}}))$, and $(\bar{A})^T = ((\overline{a_{ji}})) = (\overline{A^T})$. Hence, the transpose of the conjugate of a matrix is equal to the conjugate of the transpose of the matrix.

A matrix A such that $A = A^*$ is called a **Hermitian matrix**; that is, a matrix $A = ((a_{ij}))$ is a Hermitian matrix if and only if $a_{ij} = \overline{a_{ji}}$ for all pairs (i, j). Since $a_{ii} = \overline{a_{ii}}$ only if a_{ii} is a real number, the diagonal elements of a Hermitian matrix are real numbers. If A is a real symmetric matrix, then $a_{ij} = a_{ji}$, and $a_{ij} = \overline{a_{ji}}$ for all pairs (i, j) since $a_{ji} = \overline{a_{ji}}$. Hence, every real symmetric matrix is a Hermitian matrix.

A matrix A such that $A = -A^*$ is called a **skew-Hermitian matrix**; that is, a matrix $A = ((a_{ij}))$ is a skew-Hermitian matrix if and only if $a_{ij} = -\overline{a_{ji}}$ for all pairs (i, j). Every real skew-symmetric matrix is a skew-Hermitian matrix.

Example 1 Prove that A is a skew-Hermitian matrix where

$$A = \begin{pmatrix} 2i & 3 & i \\ -3 & 0 & -2-i \\ i & 2-i & i \end{pmatrix}.$$

$$\bar{A} = \begin{pmatrix} -2i & 3 & -i \\ -3 & 0 & -2+i \\ -i & 2+i & -i \end{pmatrix}; \quad A^* = (\bar{A})^T = \begin{pmatrix} -2i & -3 & -i \\ 3 & 0 & 2+i \\ -i & -2+i & -i \end{pmatrix};$$

$$-A^* = \begin{pmatrix} 2i & 3 & i \\ -3 & 0 & -2-i \\ i & 2-i & i \end{pmatrix} = A.$$

Hence, A is a skew-Hermitian matrix.

Example 2 Prove that $\overline{AB} = \bar{A}\bar{B}$.

Let $A = ((a_{ij}))$ and $B = ((b_{ij}))$ be matrices of order m by n and n by p, respectively. Then the ijth element of AB is given by

$$\sum_{k=1}^{n} a_{ik}b_{kj},$$

and the ijth element of $\overline{AB}$ is given by

$$\sum_{k=1}^{n} \overline{a_{ik}b_{kj}}$$

since the conjugate of the sum of complex numbers is equal to the sum of their conjugates. Also, since the conjugate of the product of two complex numbers is equal to the product of their conjugates,

$$\sum_{k=1}^{n} \overline{a_{ik}b_{kj}} = \sum_{k=1}^{n} \overline{a_{ik}}\,\overline{b_{kj}}.$$

Hence, $\overline{AB} = \bar{A}\bar{B}$; that is, the conjugate of the product of two matrices is equal to the product of their conjugates.

Example 3 Prove that $(AB)^* = B^*A^*$.

$$\begin{aligned}
(AB)^* &= (\overline{AB})^T && \text{by definition} \\
&= (\bar{A}\bar{B})^T && \text{by the results of Example 2} \\
&= (\bar{B})^T(\bar{A})^T && \text{by Theorem 1-5} \\
&= B^*A^* && \text{by definition.}
\end{aligned}$$

Additional properties of Hermitian matrices and skew-Hermitian matrices are presented in the exercises.

Exercises

1. If

$$A = \begin{pmatrix} 1 & 3 + 2i \\ -i & 2 - i \end{pmatrix},$$

find **(a)** $\bar{A}$; **(b)** A^*.

2. Determine which of the following matrices are
   (a) Hermitian matrices;   **(b)** skew-Hermitian matrices:

$$A = \begin{pmatrix} 0 & -i \\ i & 0 \end{pmatrix}, \qquad B = \begin{pmatrix} 2 & 3 + i \\ -3 - i & 5 \end{pmatrix}, \qquad C = \begin{pmatrix} 0 & 0 \\ 0 & 0 \end{pmatrix},$$

$$D = \begin{pmatrix} 3 & 5 + 2i \\ 5 - 2i & 1 \end{pmatrix}, \quad E = \begin{pmatrix} 0 & 4 + 5i \\ -4 + 5i & 0 \end{pmatrix}, \quad F = \begin{pmatrix} 2 & 3 \\ -3 & 2 \end{pmatrix},$$

$$G = \begin{pmatrix} 0 & 2 + i & i \\ 2 + i & 0 & 1 + 3i \\ i & 1 + 3i & 0 \end{pmatrix}, \quad H = \begin{pmatrix} 0 & 3 & 2 - i \\ -3 & 0 & -2i \\ -2 - i & -2i & 0 \end{pmatrix}.$$

3. Prove that the diagonal elements of a skew-Hermitian matrix are either zeros or pure imaginary numbers.

Prove the following theorems for complex matrices.

4. $(A^*)^* = A$.

5. $(A + B)^* = A^* + B^*$.

6. If k is any complex number, then $(kA)^* = \bar{k}A^*$.

7. The product of any matrix and its transposed conjugate is a Hermitian matrix; that is, $AA^* = (AA^*)^*$.

8. The sum of any square matrix and its transposed conjugate is a Hermitian matrix; that is, $A + A^* = (A + A^*)^*$.

9. The difference of any matrix and its transposed conjugate is a skew-Hermitian matrix; that is, $A - A^* = -(A - A^*)^*$.

10. Every square matrix can be expressed as the sum of a Hermitian matrix and a skew-Hermitian matrix. (Hint: See Exercises 8 and 9.)

11. Every Hermitian matrix can be expressed as $A + Bi$ where A is a real symmetric matrix and B is a real skew-symmetric matrix.

12. Every skew-Hermitian matrix can be expressed as $A + Bi$ where A is a real skew-symmetric matrix and B is a real symmetric matrix.

chapter 2

Inverses and Systems
of Matrices

2-1 Determinants

Associated with each square matrix $A = ((a_{ij}))$ of order two is a function of the matrix called its **determinant** and denoted either by det A or by

$$\begin{vmatrix} a_{11} & a_{12} \\ a_{21} & a_{22} \end{vmatrix}.$$

The determinant function assigns to each square real matrix A of order two a unique real number $a_{11}a_{22} - a_{21}a_{12}$ called the **value of the determinant**; that is, det $A = a_{11}a_{22} - a_{21}a_{12}$. The value of the determinant of a matrix of order two may be remembered by the array

$$\begin{vmatrix} a_{11} & a_{12} \\ a_{21} & a_{22} \end{vmatrix} \begin{matrix} - \\ + \end{matrix} = a_{11}a_{22} - a_{21}a_{12}. \tag{2-1}$$

Note that the value of the determinant is the difference of the product of the diagonal elements and the product of the remaining two elements.

Example 1 Find the value of the determinant of matrix A

where
$$A = \begin{pmatrix} 2 & -1 \\ 4 & 3 \end{pmatrix}.$$

$$\det A = \begin{vmatrix} 2 & -1 \\ 4 & 3 \end{vmatrix} = (2)(3) - (4)(-1) = 10.$$

A **determinant** of a matrix $A = ((a_{ij}))$ of order three is a function of the square matrix and is denoted by det A or by

$$\begin{vmatrix} a_{11} & a_{12} & a_{13} \\ a_{21} & a_{22} & a_{23} \\ a_{31} & a_{32} & a_{33} \end{vmatrix}.$$

The value of the determinant of A is defined as

$$a_{11}a_{22}a_{33} + a_{12}a_{23}a_{31} + a_{13}a_{21}a_{32}$$
$$- a_{31}a_{22}a_{13} - a_{32}a_{23}a_{11} - a_{33}a_{21}a_{12}.$$

The value of the determinant of a matrix of order three may be remembered by use of a particular scheme similar to that used for determinants of matrices of order two (2-1):

$$- a_{31}a_{22}a_{13} - a_{32}a_{23}a_{11} - a_{33}a_{21}a_{12} \qquad ,$$

$$(2\text{-}2)$$

$$+ a_{11}a_{22}a_{33} + a_{12}a_{23}a_{31} + a_{13}a_{21}a_{32}$$

In (2-2) the first two columns of the determinant are repeated at its right. The products of the three elements along the arrows running downward and to the right are noted as well as the negative of the products of the three elements along the arrows running upward and to the right. The algebraic sum of these six products is the value of the determinant.

Example 2 Find the value of the determinant of matrix A where

$$A = \begin{pmatrix} 0 & 4 & 2 \\ 4 & -2 & -1 \\ 5 & 1 & 3 \end{pmatrix}.$$

Using the scheme of (2-2),

$$-(-20) - (0) - (48)$$

$$+ (0) + (-20) + (8)$$

Hence, the value of the determinant of matrix A is $(0) + (-20) + (8) - (-20) - (0) - (48)$; that is, det $A = -40$.

Note carefully that the method of diagonals employed for the evaluation of determinants of matrices of order three is not valid for determinants of matrices of higher order.

In general, a **determinant** of a matrix $A = ((a_{ij}))$ of order n is a function of the square matrix and is denoted by det A or by

$$\begin{vmatrix} a_{11} & a_{12} & \cdots & a_{1n} \\ a_{21} & a_{22} & \cdots & a_{2n} \\ \cdots & \cdots & \cdots & \cdots \\ a_{n1} & a_{n2} & \cdots & a_{nn} \end{vmatrix}.$$

The value of the determinant of a matrix of order n is defined as the sum of $n!$ terms of the form

$$(-1)^k a_{1i_1} a_{2i_2} \cdots a_{ni_n}.$$

Each term contains one and only one element from each row and one and only one element from each column; that is, the second subscripts $i_1, i_2, \ldots, i_n$ are equal to $1, 2, \ldots, n$, taken in some order. The exponent k represents the number of interchanges of two elements necessary for the second subscripts to be placed in the order $1, 2, \ldots, n$. For example, consider the term containing $a_{13} a_{21} a_{34} a_{42}$ in the evaluation of the determinant of a matrix of order four. The value of k is 3 since three interchanges of two elements are necessary for the second subscripts to be placed in the order $1, 2, 3, 4$:

$$a_{13} a_{21} a_{34} a_{42} = a_{21} a_{13} a_{34} a_{42} = a_{21} a_{42} a_{34} a_{13} = a_{21} a_{42} a_{13} a_{34}.$$

Hence, the term containing the factor $a_{13} a_{21} a_{34} a_{42}$ has the additional factor $(-1)^3$; that is, -1.

Although the following theorems are valid for determinants of matrices of order n, their proofs shall be considered only for determinants of matrices of order three. These theorems are extremely useful in the evaluation of determinants of matrices of order $n \geq 4$ since the values of such determinants are seldom computed from the definition. In addition, several of these theorems shall be used to discuss certain concepts in matrix algebra.

Theorem 2-1 *The value of a determinant remains unchanged if corresponding rows and columns are interchanged; that is, det $A = det\ A^T$.*

   Proof: Consider the determinants

$$\det A = \begin{vmatrix} a_{11} & a_{12} & a_{13} \\ a_{21} & a_{22} & a_{23} \\ a_{31} & a_{32} & a_{33} \end{vmatrix} \quad \text{and} \quad \det A^T = \begin{vmatrix} a_{11} & a_{21} & a_{31} \\ a_{12} & a_{22} & a_{32} \\ a_{13} & a_{23} & a_{33} \end{vmatrix}.$$

Note that the elements of each row or column of det A correspond to the elements of the same numbered column or row, respectively, of det A^T. By definition,

$$\det A = a_{11} a_{22} a_{33} + a_{12} a_{23} a_{31} + a_{13} a_{21} a_{32}$$
$$- a_{31} a_{22} a_{13} - a_{32} a_{23} a_{11} - a_{33} a_{21} a_{12}$$

and

$$\det A^T = a_{11} a_{22} a_{33} + a_{21} a_{32} a_{13} + a_{31} a_{12} a_{23}$$
$$- a_{13} a_{22} a_{31} - a_{23} a_{32} a_{11} - a_{33} a_{12} a_{21}.$$

By a simple rearrangement of factors and terms, note that
$$\det A = \det A^T.$$

The importance of Theorem 2-1 is that for every proved theorem concerning the rows of a determinant there exists a corresponding theorem concerning the columns. The corresponding theorem will not be stated in each instance, but the reader should interpret each theorem in terms of the corresponding theorem as well.

Theorem 2-2 *If any two rows of a determinant are interchanged, then the sign of the value of the determinant is changed.*

For example,

$$\begin{vmatrix} a_{11} & a_{12} & a_{13} \\ a_{31} & a_{32} & a_{33} \\ a_{21} & a_{22} & a_{23} \end{vmatrix} = - \begin{vmatrix} a_{11} & a_{12} & a_{13} \\ a_{21} & a_{22} & a_{23} \\ a_{31} & a_{32} & a_{33} \end{vmatrix}.$$

The proof of Theorem 2-2 is left to the reader as an exercise.

Theorem 2-3 *If every element of a row of a determinant is zero, then the value of the determinant is zero.*

Proof: Consider the definition of the value of the determinant of a matrix of order three. Every term of this expression contains one and only one element from each row as a factor. Therefore, every term must contain a factor which is an element from the row of zeros. Hence, the value of the determinant is zero.

Theorem 2-4 *If every element of a row of a determinant is multiplied by the factor k, then the value of the determinant is multiplied by k.*

For example,

$$\begin{vmatrix} ka_{11} & ka_{12} & ka_{13} \\ a_{21} & a_{22} & a_{23} \\ a_{31} & a_{32} & a_{33} \end{vmatrix} = k \begin{vmatrix} a_{11} & a_{12} & a_{13} \\ a_{21} & a_{22} & a_{23} \\ a_{31} & a_{32} & a_{33} \end{vmatrix}.$$

Proof: We have already mentioned that every term in the definition of the value of the determinant of a matrix of order three contains one and only one element from each row as a factor. Therefore, every term will contain the factor k once and only once; that is, the value of the determinant will be multiplied by k.

Theorem 2-5 *The value of a determinant remains unchanged if every element of a row is increased by a scalar multiple of the corresponding element of another row.*

For example,

$$\begin{vmatrix} a_{11} & a_{12} & a_{13} \\ a_{21} + ka_{31} & a_{22} + ka_{32} & a_{23} + ka_{33} \\ a_{31} & a_{32} & a_{33} \end{vmatrix} = \begin{vmatrix} a_{11} & a_{12} & a_{13} \\ a_{21} & a_{22} & a_{23} \\ a_{31} & a_{32} & a_{33} \end{vmatrix}.$$

The reader should be careful to note that the elements of the third row are left intact while a scalar multiple of these elements is added to each of the corresponding elements of the second row. The proof of Theorem 2-5 may be obtained by use of the definition of the value of the determinant of a matrix of order three and is left to the reader as an exercise.

The value of the determinant of a matrix of order three has been defined by the equation

$$\begin{vmatrix} a_{11} & a_{12} & a_{13} \\ a_{21} & a_{22} & a_{23} \\ a_{31} & a_{32} & a_{33} \end{vmatrix} = \begin{aligned} & a_{11}a_{22}a_{33} + a_{12}a_{23}a_{31} + a_{13}a_{21}a_{32} \\ & - a_{31}a_{22}a_{12} - a_{32}a_{23}a_{11} - a_{33}a_{21}a_{12}. \end{aligned} \tag{2-3}$$

By a rearrangement of terms, the right-hand member of (2-3) may be written as

$$a_{11}(a_{22}a_{33} - a_{32}a_{23}) - a_{12}(a_{21}a_{33} - a_{31}a_{23}) + a_{13}(a_{21}a_{32} - a_{31}a_{22}),$$

and

$$a_{11} \begin{vmatrix} a_{22} & a_{23} \\ a_{32} & a_{33} \end{vmatrix} - a_{12} \begin{vmatrix} a_{21} & a_{23} \\ a_{31} & a_{33} \end{vmatrix} + a_{13} \begin{vmatrix} a_{21} & a_{22} \\ a_{31} & a_{32} \end{vmatrix}. \tag{2-4}$$

The three determinants in (2-4) may be obtained from the original determinant by eliminating certain rows and columns. The pattern shown by (2-4) for expressing the value of the determinant of a matrix of order three may be generalized for determinants of matrices of higher order.

Consider the determinant of a square matrix $((a_{ij}))$ of order n. The determinant obtained from det $((a_{ij}))$ by deleting the elements of the ith row and jth column is called the **minor** of the element a_{ij} and is denoted by M_{ij}. The minor, when prefixed by a positive or negative sign according to whether the sum of the position numbers of the row and column deleted from det $((a_{ij}))$ is even or odd, respectively, is called the **cofactor** of the element a_{ij} and is denoted by the symbol A_{ij}. That is,

$$A_{ij} = (-1)^{i+j}M_{ij}. \tag{2-5}$$

For example, the minor of a_{12} in

$$\det A = \begin{vmatrix} 2 & 1 & 3 \\ 4 & -1 & 2 \\ -2 & 0 & 1 \end{vmatrix}$$

is

$$M_{12} = \begin{vmatrix} 4 & 2 \\ -2 & 1 \end{vmatrix} = (4)(1) - (-2)(2) = 4 + 4 = 8$$

and the cofactor is

$$A_{12} = (-1)^{1+2}M_{12} = (-1)(8) = -8.$$

Theorem 2-6 *The value of a determinant is equal to the sum of the products of each element of any row and its cofactor.*

Theorem 2-6 is a key theorem for evaluating determinants of matrices of order n. The proof of Theorem 2-6 for determinants of matrices of order n may be found in most linear algebra texts.

Example 3 Evaluate

$$\begin{vmatrix} 2 & -3 & 1 \\ 0 & 5 & 2 \\ -1 & -2 & 3 \end{vmatrix}$$

using the cofactors of the elements of the second column.

The cofactors of -3, 5, and -2, the elements of the second column, are denoted symbolically by A_{12}, A_{22}, and A_{32}, respectively. Now,

$$A_{12} = (-1)^{1+2}\begin{vmatrix} 0 & 2 \\ -1 & 3 \end{vmatrix} = -2; \quad A_{22} = (-1)^{2+2}\begin{vmatrix} 2 & 1 \\ -1 & 3 \end{vmatrix} = 7;$$

$$A_{32} = (-1)^{3+2}\begin{vmatrix} 2 & 1 \\ 0 & 2 \end{vmatrix} = -4.$$

Using Theorems 2-1 and 2-6,

$$\begin{vmatrix} 2 & -3 & 1 \\ 0 & 5 & 2 \\ -1 & -2 & 3 \end{vmatrix} = (-3)A_{12} + (5)A_{22} + (-2)A_{32}$$

$$= (-3)(-2) + (5)(7) + (-2)(-4) = 49.$$

The symbol $|A|$ is also used to denote the determinant of A. Be careful to note that $|A|$ does *not* mean the absolute value of A.

Exercises

In Exercises 1 through 6 find the value of the given determinant.

1. $\begin{vmatrix} 5 & 3 \\ 2 & 4 \end{vmatrix}$.

2. $\begin{vmatrix} 4 & -2 \\ 2 & 1 \end{vmatrix}$.

3. $\begin{vmatrix} 5 & 2 \\ 1 & 0 \end{vmatrix}$.

4. $\begin{vmatrix} 8 & 4 \\ 2 & 1 \end{vmatrix}$.

5. $\begin{vmatrix} 2 & 3 & 1 \\ 1 & 4 & -3 \\ -1 & 2 & 0 \end{vmatrix}$.

6. $\begin{vmatrix} -3 & 1 & 1 \\ 1 & -3 & 1 \\ 1 & 1 & -3 \end{vmatrix}$.

7. Prove that the value of the determinant of a matrix of order n is zero if two rows are identical.

8. Prove that for any determinant of a matrix of order three the sum of the products of the elements of any row and the cofactors of the corresponding elements of another row is zero. (Note: This result is true for the determinant of a matrix of any order n.)

9. Find det $((ka_{ij}))$ if $((a_{ij}))$ is of order four and det $((a_{ij})) = m$.

10. Prove that det $AB = \det A \det B$ for any two square matrices A and B of order two. (Note: This result is true for square matrices A and B of any order n.)

11. Verify the results of Exercise 10 where

$$A = \begin{pmatrix} 3 & 2 \\ 5 & 1 \end{pmatrix} \quad \text{and} \quad B = \begin{pmatrix} 1 & 6 \\ 2 & 9 \end{pmatrix}.$$

12. Matrix $((a_{ij}))$ of order n is called an **upper triangular matrix** if $a_{ij} = 0$ for every pair (i, j) such that $i > j$. Find det $((a_{ij}))$ for such a matrix.

13. Show that the equation

$$\begin{vmatrix} 5 - \lambda & 1 \\ 2 & 3 - \lambda \end{vmatrix} = 0$$

is satisfied if λ is replaced by the matrix

$$A = \begin{pmatrix} 5 & 1 \\ 2 & 3 \end{pmatrix},$$

and each real number n is replaced by the scalar multiple nI where I is the identity matrix of order two.

14. Prove that

$$\begin{vmatrix} a & b & m & n \\ c & d & r & s \\ 0 & 0 & e & f \\ 0 & 0 & g & h \end{vmatrix} = \begin{vmatrix} a & b \\ c & d \end{vmatrix} \begin{vmatrix} e & f \\ g & h \end{vmatrix}.$$

15. Prove that

$$\begin{vmatrix} a + b & a & a \\ a & a + b & a \\ a & a & a + b \end{vmatrix} = b^2(3a + b).$$

16. Repeat Exercise 6 using the formula of Exercise 15.

2-2 Inverse of a Matrix

This section will be concerned with the problem of finding a multiplicative inverse, if it exists, for any given square matrix. A **left multiplicative inverse** of a matrix A is a matrix B such that $BA = I$; a **right multiplicative inverse** of a matrix A is a matrix C such that $AC = I$. If a left and a right multiplicative inverse of a matrix A are equal, then the left (right) inverse is called, simply, a **multiplicative inverse** of A and is denoted by A^{-1}.

Theorem 2-7 *A left multiplicative inverse of a square matrix A is a multiplicative inverse of A.*

> *Proof:* Suppose $BA = I$, then
> $A(BA) = AI$ by premultiplication by A,
> $(AB)A = A$ since the multiplication of matrices
>   is associative and $AI = A$,
> $AB = I$ since I is the unique matrix such that $IA = A$.
> Therefore, B is also a right multiplicative inverse of A. Hence, B is a multiplicative inverse of A; that is, $B = A^{-1}$.

Similarly, it can be shown that Theorem 2-8 is true. The proof is left to the reader as an exercise.

Theorem 2-8 *A right multiplicative inverse of a square matrix A is a multiplicative inverse of A.*

Theorem 2-9 *The multiplicative inverse, if it exists, of a square matrix A is unique.*

> *Proof:* Let A^{-1} and B be any two multiplicative inverses of the square matrix A. Since $A^{-1}A = I$ and $BA = I$, then
>
> $$A^{-1}A = BA$$
> $$(A^{-1}A)A^{-1} = (BA)A^{-1}$$
> $$A^{-1}(A A^{-1}) = B(A A^{-1})$$
> $$A^{-1}I = BI$$
> $$A^{-1} = B.$$

Hence, any two multiplicative inverses of matrix A are identically equal; that is, the multiplicative inverse, if it exists, of a square matrix is unique.

Not every square matrix has a multiplicative inverse. For example, the matrix

$$\begin{pmatrix} 1 & 2 \\ 0 & 0 \end{pmatrix}$$

does not have a multiplicative inverse. If

$$\begin{pmatrix} a & b \\ c & d \end{pmatrix}$$

were the multiplicative inverse of

$$\begin{pmatrix} 1 & 2 \\ 0 & 0 \end{pmatrix},$$

then

$$\begin{pmatrix} a & b \\ c & d \end{pmatrix}\begin{pmatrix} 1 & 2 \\ 0 & 0 \end{pmatrix} = \begin{pmatrix} 1 & 0 \\ 0 & 1 \end{pmatrix};$$

$$\begin{pmatrix} a & 2a \\ c & 2c \end{pmatrix} = \begin{pmatrix} 1 & 0 \\ 0 & 1 \end{pmatrix}. \tag{2-6}$$

This requires that four equations be satisfied: $a = 1$, $2a = 0$, $c = 0$, $2c = 1$. However, if $a = 1$, then $2a \neq 0$. Hence, values of a, b, c, and d which satisfy the matrix equation (2-6) do not exist, and the matrix

$$\begin{pmatrix} 1 & 2 \\ 0 & 0 \end{pmatrix}$$

does not have a multiplicative inverse.

Example 1 Find the multiplicative inverse, if it exists, of

$$\begin{pmatrix} 3 & 1 \\ 4 & 2 \end{pmatrix}.$$

If

$$\begin{pmatrix} 3 & 1 \\ 4 & 2 \end{pmatrix}$$

has a multiplicative inverse

$$\begin{pmatrix} a & b \\ c & d \end{pmatrix},$$

then

$$\begin{pmatrix} a & b \\ c & d \end{pmatrix}\begin{pmatrix} 3 & 1 \\ 4 & 2 \end{pmatrix} = \begin{pmatrix} 1 & 0 \\ 0 & 1 \end{pmatrix};$$

$$\begin{pmatrix} 3a + 4b & a + 2b \\ 3c + 4d & c + 2d \end{pmatrix} = \begin{pmatrix} 1 & 0 \\ 0 & 1 \end{pmatrix}.$$

Therefore,

$$\begin{cases} 3a + 4b = 1 \\ a + 2b = 0 \end{cases} \quad \text{and} \quad \begin{cases} 3c + 4d = 0 \\ c + 2d = 1 \end{cases}.$$

Solve these pairs of equations simultaneously and obtain $a = 1$, $b = -\frac{1}{2}$, $c = -2$, and $d = \frac{3}{2}$. Hence, the multiplicative inverse of

$$\begin{pmatrix} 3 & 1 \\ 4 & 2 \end{pmatrix} \quad \text{is} \quad \begin{pmatrix} 1 & -\frac{1}{2} \\ -2 & \frac{3}{2} \end{pmatrix}.$$

Example 2 Find the form of the multiplicative inverse, if it exists, of the general square matrix of order two

$$\begin{pmatrix} a & b \\ c & d \end{pmatrix}.$$

If

$$\begin{pmatrix} a & b \\ c & d \end{pmatrix}$$

has a multiplicative inverse

$$\begin{pmatrix} w & x \\ y & z \end{pmatrix},$$

then

$$\begin{pmatrix} w & x \\ y & z \end{pmatrix}\begin{pmatrix} a & b \\ c & d \end{pmatrix} = \begin{pmatrix} 1 & 0 \\ 0 & 1 \end{pmatrix};$$

$$\begin{pmatrix} aw + cx & bw + dx \\ ay + cz & by + dz \end{pmatrix} = \begin{pmatrix} 1 & 0 \\ 0 & 1 \end{pmatrix}.$$

Therefore,

$$\begin{cases} aw + cx = 1 \\ bw + dx = 0 \end{cases} \text{ and } \begin{cases} ay + cz = 0 \\ by + dz = 1 \end{cases}.$$

If $ad - bc \neq 0$, the solutions of these pairs of equations are given as

$$w = \frac{d}{ad - bc}, \quad x = \frac{-b}{ad - bc},$$

$$y = \frac{-c}{ad - bc}, \quad z = \frac{a}{ad - bc}.$$

Hence, the multiplicative inverse of

$$\begin{pmatrix} a & b \\ c & d \end{pmatrix} \text{ is } \begin{pmatrix} \dfrac{d}{ad - bc} & \dfrac{-b}{ad - bc} \\ \dfrac{-c}{ad - bc} & \dfrac{a}{ad - bc} \end{pmatrix}$$

providing $ad - bc \neq 0$. Notice that the multiplicative inverse of the square matrix of order two

$$\begin{pmatrix} a & b \\ c & d \end{pmatrix}$$

exists if

$$\begin{vmatrix} a & b \\ c & d \end{vmatrix} \neq 0.$$

Examples 1 and 2 illustrate the use of a method that involves considerable labor and is impractical for finding the multiplicative inverses of square matrices of orders greater than two. Now, a direct method for finding the multiplicative inverse, if it exists, of a square matrix of any order will be derived. If $A = ((a_{ij}))$, then by Theorem 2-6

$$\det A = \sum_{j=1}^{n} a_{ij} A_{ij}, \quad \text{for} \quad i = 1, 2, \ldots, n. \tag{2-7}$$

By Exercise 8 of §2-1,

$$\sum_{j=1}^{n} a_{hj} A_{ij} = 0, \quad \text{for} \quad h, i = 1, 2, \ldots, n \quad \text{where} \quad h \neq i. \tag{2-8}$$

Equations (2-7) and (2-8) may be expressed as the single equation

$$\sum_{j=1}^{n} a_{hj} A_{ij} = \delta_{hi} \det A, \quad \text{for} \quad h, i = 1, 2, \ldots, n. \tag{2-9}$$

The n^2 equations represented by (2-9) may be written in matrix form as

$$((a_{ij}))((A_{ij}))^T = ((\delta_{ij})) \det A. \tag{2-10}$$

For example, if $n = 3$ in equation (2-9), then equation (2-10) represents

$$\begin{pmatrix} a_{11} & a_{12} & a_{13} \\ a_{21} & a_{22} & a_{23} \\ a_{31} & a_{32} & a_{33} \end{pmatrix} \begin{pmatrix} A_{11} & A_{21} & A_{31} \\ A_{12} & A_{22} & A_{32} \\ A_{13} & A_{23} & A_{33} \end{pmatrix} = \begin{pmatrix} 1 & 0 & 0 \\ 0 & 1 & 0 \\ 0 & 0 & 1 \end{pmatrix} \det A.$$

Since $((a_{ij})) = A$ and $((\delta_{ij})) = I$, equation (2-10) may be written as

$$A ((A_{ij}))^T = I \det A. \tag{2-11}$$

If $\det A \neq 0$, then

$$A \frac{((A_{ij}))^T}{\det A} = I. \tag{2-12}$$

Hence, the matrix $\dfrac{((A_{ij}))^T}{\det A}$ is the multiplicative inverse of A; that is,

$$A^{-1} = \frac{((A_{ij}))^T}{\det A} = \frac{1}{\det A} ((A_{ij}))^T. \tag{2-13}$$

The multiplicative inverse, if it exists, of a square matrix is the product of the reciprocal of the determinant of that matrix and the transpose of the **matrix of cofactors**. That $\dfrac{1}{\det A}((A_{ij}))^T$ is not just the right inverse of A follows from Theorem 2-8.

Note that a necessary and sufficient condition for the multiplicative inverse of matrix A to exist is that $\det A \neq 0$. A square matrix A is said to be **nonsingular** if $\det A \neq 0$, and **singular** if $\det A = 0$.

It should be mentioned that if A is not a square matrix, then it is possible for A to have a left or a right multiplicative inverse, but not both. For example, consider

$$A = \begin{pmatrix} 1 & 0 & 0 \\ 0 & 1 & 0 \end{pmatrix}.$$

Then any matrix of the form

$$\begin{pmatrix} 1 & 0 \\ 0 & 1 \\ r & s \end{pmatrix},$$

where r and s are arbitrary scalars, is a right multiplicative inverse of A. A left multiplicative inverse

$$\begin{pmatrix} m & n \\ w & x \\ y & z \end{pmatrix}$$

does not exist since

$$\begin{pmatrix} m & n \\ w & x \\ y & z \end{pmatrix} \begin{pmatrix} 1 & 0 & 0 \\ 0 & 1 & 0 \end{pmatrix} = \begin{pmatrix} m & n & 0 \\ w & x & 0 \\ y & z & 0 \end{pmatrix} \neq \begin{pmatrix} 1 & 0 & 0 \\ 0 & 1 & 0 \\ 0 & 0 & 1 \end{pmatrix}$$

for any values of m, n, w, x, y, and z.

Example 3 Find the multiplicative inverse, if it exists, of A where

$$A = \begin{pmatrix} 1 & 2 & 3 \\ 1 & 3 & 5 \\ 1 & 5 & 12 \end{pmatrix}.$$

Using Theorem 2-6 and the cofactors of the elements of the first row,

$$\det A = \begin{vmatrix} 3 & 5 \\ 5 & 12 \end{vmatrix} - 2 \begin{vmatrix} 1 & 5 \\ 1 & 12 \end{vmatrix} + 3 \begin{vmatrix} 1 & 3 \\ 1 & 5 \end{vmatrix} = 11 - 2(7) + 3(2) = 3.$$

Since $\det A \neq 0$, then A^{-1} exists. The cofactors of the elements of A are given by

$$A_{11} = \begin{vmatrix} 3 & 5 \\ 5 & 12 \end{vmatrix} = 11, \qquad A_{12} = -\begin{vmatrix} 1 & 5 \\ 1 & 12 \end{vmatrix} = -7, \quad A_{13} = \begin{vmatrix} 1 & 3 \\ 1 & 5 \end{vmatrix} = 2,$$

$$A_{21} = -\begin{vmatrix} 2 & 3 \\ 5 & 12 \end{vmatrix} = -9, \quad A_{22} = \begin{vmatrix} 1 & 3 \\ 1 & 12 \end{vmatrix} = 9, \qquad A_{23} = -\begin{vmatrix} 1 & 2 \\ 1 & 5 \end{vmatrix} = -3,$$

$$A_{31} = \begin{vmatrix} 2 & 3 \\ 3 & 5 \end{vmatrix} = 1, \qquad A_{32} = -\begin{vmatrix} 1 & 3 \\ 1 & 5 \end{vmatrix} = -2, \quad A_{33} = \begin{vmatrix} 1 & 2 \\ 1 & 3 \end{vmatrix} = 1.$$

Replacing each element of A by its cofactor, we obtain the matrix

$$((A_{ij})) = \begin{pmatrix} 11 & -7 & 2 \\ -9 & 9 & -3 \\ 1 & -2 & 1 \end{pmatrix}.$$

Hence,

$$A^{-1} = \frac{((A_{ij}))^T}{\det A} = \frac{1}{3}\begin{pmatrix} 11 & -9 & 1 \\ -7 & 9 & -2 \\ 2 & -3 & 1 \end{pmatrix} = \begin{pmatrix} \frac{11}{3} & -3 & \frac{1}{3} \\ -\frac{7}{3} & 3 & -\frac{2}{3} \\ \frac{2}{3} & -1 & \frac{1}{3} \end{pmatrix}.$$

Example 4 Solve the system of linear equations

$$\begin{cases} 5x - 2y = 12 \\ x + 2y = 0. \end{cases}$$

This system of linear equations may be expressed in matrix form as

$$\begin{pmatrix} 5 & -2 \\ 1 & 2 \end{pmatrix}\begin{pmatrix} x \\ y \end{pmatrix} = \begin{pmatrix} 12 \\ 0 \end{pmatrix}.$$

The multiplicative inverse of

$$\begin{pmatrix} 5 & -2 \\ 1 & 2 \end{pmatrix} \quad \text{is} \quad \begin{pmatrix} \frac{1}{6} & \frac{1}{6} \\ -\frac{1}{12} & \frac{5}{12} \end{pmatrix}.$$

Premultiply both sides of the matrix equation representing the system of linear equations by this multiplicative inverse to obtain

$$\begin{pmatrix} \frac{1}{6} & \frac{1}{6} \\ -\frac{1}{12} & \frac{5}{12} \end{pmatrix}\begin{pmatrix} 5 & -2 \\ 1 & 2 \end{pmatrix}\begin{pmatrix} x \\ y \end{pmatrix} = \begin{pmatrix} \frac{1}{6} & \frac{1}{6} \\ -\frac{1}{12} & \frac{5}{12} \end{pmatrix}\begin{pmatrix} 12 \\ 0 \end{pmatrix}$$

$$\begin{pmatrix} 1 & 0 \\ 0 & 1 \end{pmatrix}\begin{pmatrix} x \\ y \end{pmatrix} = \begin{pmatrix} 2 \\ -1 \end{pmatrix}$$

$$\begin{pmatrix} x \\ y \end{pmatrix} = \begin{pmatrix} 2 \\ -1 \end{pmatrix}.$$

Hence, $x = 2$ and $y = -1$.

In subsequent sections other methods of determining the multiplicative inverse of a nonsingular matrix will be examined.

Exercises

In Exercises 1 through 6 determine the multiplicative inverse, if it exists, of the given matrix.

1. $\begin{pmatrix} 3 & 2 \\ 5 & 4 \end{pmatrix}.$

2. $\begin{pmatrix} 3 & 6 \\ 1 & 2 \end{pmatrix}.$

3. $\begin{pmatrix} \cos\theta & -\sin\theta \\ \sin\theta & \cos\theta \end{pmatrix}.$

4. $\begin{pmatrix} 2 & 3 & 4 \\ 0 & 5 & 6 \\ 0 & 0 & 1 \end{pmatrix}.$

5. $\begin{pmatrix} 1 & 2 & 3 \\ 2 & 4 & 5 \\ 3 & 5 & 6 \end{pmatrix}.$

6. $\begin{pmatrix} 0 & a & b \\ -a & 0 & c \\ -b & -c & 0 \end{pmatrix}.$

In Exercises 7 and 8 solve the system of simultaneous equations by matrix methods.

7. $\begin{cases} 2x + y = 4 \\ 3x + 4y = 1. \end{cases}$

8. $\begin{cases} x + 3y + 3z = 2 \\ x + 3y + 4z = 3 \\ x + 4y + 3z = 1. \end{cases}$

9. Find a left multiplicative inverse of

$$\begin{pmatrix} 1 & 1 \\ 3 & 4 \\ 0 & 0 \end{pmatrix}.$$

Show that a right multiplicative inverse does not exist.

10. Prove that $(A^{-1})^{-1} = A$.

11. Prove that if $AB = 0$ and $B \neq 0$, then A is a singular matrix.

12. Prove that $(AB)^{-1} = B^{-1}A^{-1}$.

13. Verify the results of Exercise 12 for

$$A = \begin{pmatrix} 6 & 2 \\ 5 & 2 \end{pmatrix} \quad \text{and} \quad B = \begin{pmatrix} 1 & 0 \\ -1 & 2 \end{pmatrix}.$$

14. Prove that if $AB = AC$ and A is a nonsingular matrix, then $B = C$.

15. Prove that the multiplicative inverse of a nonsingular symmetric matrix is a symmetric matrix.

16. Prove that if $AB = BA$, then $A^{-1}B^{-1} = B^{-1}A^{-1}$.

17. Prove that $(A^T)^{-1} = (A^{-1})^T$ for any nonsingular matrix A.

18. Verify the results of Exercise 17 for

$$A = \begin{pmatrix} 6 & 3 \\ 8 & 6 \end{pmatrix}.$$

2-3 Systems of Matrices

The set of square matrices of a given order forms, under the operations of matrix addition and multiplication, a model of a particular abstract mathematical system called a ring. A **ring** is a mathematical system consisting of a set R of elements $a, b, c, \ldots$, an equivalence relation denoted by $=$, and two well-defined binary operations $+$ and $\times$ called "addition" and "multiplication," respectively, which satisfy the following properties:

(1) *Closure:* If $a, b \in R$, then $a + b \in R$ and $a \times b \in R$.

(2) *Commutative:* If $a, b \in R$, then $a + b = b + a$.

(3) *Associative:* If $a, b, c \in R$, then $a + (b + c) = (a + b) + c$ and $a \times (b \times c) = (a \times b) \times c$.

(4) *Additive identity:* There exists an element $z \in R$ such that for every $a \in R$, $a + z = z + a = a$.

(5) *Additive inverse:* For each $a \in R$ there exists an element $(-a) \in R$ such that $a + (-a) = (-a) + a = z$.

(6) *Distributive:* For every $a, b, c \in R$, $a \times (b + c) = (a \times b) + (a \times c)$ and $(b + c) \times a = (b \times a) + (c \times a)$.

Example 1 Show that the set S of square real matrices of order n forms a ring.

> By the definitions of the sum and product of two matrices, the set S is closed under matrix addition and multiplication (Property 1). The addition of two matrices of S is commutative (Property 2) and associative (part of Property 3) since the addition of real numbers is commutative and associative. Property 4 is satisfied because the zero matrix of order n is the additive identity element for the set S. Since for each $((a_{ij})) \in S$ there exists a matrix $((-a_{ij}))$ such that $((a_{ij})) + ((-a_{ij})) = 0$, Property 5 is satisfied. Property 6 and the second part of Property 3 were proved in §1-2. Therefore, the set S of square real matrices of order n forms a ring.

A one-to-one correspondence between the elements of a ring R and the elements of a ring S is called an **isomorphism** if sums and products of corresponding elements correspond; that is, if any two elements $a, b \in R$ and any two elements $a', b' \in S$ are such that a and b may be made to correspond to a' and b', respectively, denoted by $a \leftrightarrow a'$ and $b \leftrightarrow b'$, the correspondence is an isomorphism if $a + b \leftrightarrow a' + b'$ and $ab \leftrightarrow a'b'$. Whenever an isomorphism between the elements of a ring R and those of a ring S exists, the rings are abstractly identical and are said to be **isomorphic**. Only the notation used to represent the elements and the operations of the two rings differ.

Two very important subsets of the set of square real matrices of order two exist. The first is the set of scalar matrices of order two

$$\begin{pmatrix} k & 0 \\ 0 & k \end{pmatrix}.$$

This set of matrices is isomorphic to the set of real numbers. Both the set of scalar matrices and the set of real numbers along with the operations of addition and multiplication defined on the sets are examples of rings.

Consider the one-to-one correspondence between the matrices

$$\begin{pmatrix} k & 0 \\ 0 & k \end{pmatrix}$$

and the scalars k; that is,

$$\begin{pmatrix} k & 0 \\ 0 & k \end{pmatrix} \leftrightarrow k.$$

In order to establish that the two systems are isomorphic, it is necessary to show that the sums and products of corresponding elements in the two systems are in the same one-to-one correspondence. Let

$$\begin{pmatrix} a & 0 \\ 0 & a \end{pmatrix} \longleftrightarrow a \quad \text{and} \quad \begin{pmatrix} b & 0 \\ 0 & b \end{pmatrix} \longleftrightarrow b.$$

Then

$$\begin{pmatrix} a & 0 \\ 0 & a \end{pmatrix} + \begin{pmatrix} b & 0 \\ 0 & b \end{pmatrix} = \begin{pmatrix} a+b & 0 \\ 0 & a+b \end{pmatrix} \longleftrightarrow a+b,$$

and

$$\begin{pmatrix} a & 0 \\ 0 & a \end{pmatrix} \begin{pmatrix} b & 0 \\ 0 & b \end{pmatrix} = \begin{pmatrix} ab & 0 \\ 0 & ab \end{pmatrix} \longleftrightarrow ab.$$

Hence, the ring of matrices of the form

$$\begin{pmatrix} k & 0 \\ 0 & k \end{pmatrix}$$

is isomorphic to the ring of real numbers k.

The second important subset of the set of square real matrices of order two is the set of matrices of the form

$$\begin{pmatrix} x & y \\ -y & x \end{pmatrix}.$$

This set of matrices is isomorphic to the set of complex numbers $x + yi$. Both the set of matrices of this form and the set of complex numbers along with the operations of addition and multiplication defined on the sets are examples of rings.

Consider the one-to-one correspondence between the matrices

$$\begin{pmatrix} x & y \\ -y & x \end{pmatrix}$$

and the complex numbers $x + yi$; that is,

$$\begin{pmatrix} x & y \\ -y & x \end{pmatrix} \longleftrightarrow x + yi.$$

Let

$$\begin{pmatrix} a & b \\ -b & a \end{pmatrix} \longleftrightarrow a + bi \quad \text{and} \quad \begin{pmatrix} c & d \\ -d & c \end{pmatrix} \longleftrightarrow c + di.$$

Then

$$\begin{pmatrix} a & b \\ -b & a \end{pmatrix} + \begin{pmatrix} c & d \\ -d & c \end{pmatrix} = \begin{pmatrix} a+c & b+d \\ -(b+d) & a+c \end{pmatrix}$$

and $(a + bi) + (c + di) = (a + c) + (b + d)i$. Since

$$\begin{pmatrix} a+c & b+d \\ -(b+d) & a+c \end{pmatrix} \longleftrightarrow (a+c) + (b+d)i,$$

then

$$\begin{pmatrix} a & b \\ -b & a \end{pmatrix} + \begin{pmatrix} c & d \\ -d & c \end{pmatrix} \longleftrightarrow (a + bi) + (c + di),$$

which demonstrates that the correspondence is preserved under addition.

For multiplication,

$$\begin{pmatrix} a & b \\ -b & a \end{pmatrix} \begin{pmatrix} c & d \\ -d & c \end{pmatrix} = \begin{pmatrix} ac - bd & ad + bc \\ -(ad + bc) & ac - bd \end{pmatrix}$$

and $(a + bi) \times (c + di) = (ac - bd) + (ad + bc)i$. Since

$$\begin{pmatrix} ac - bd & ad + bc \\ -(ad + bc) & ac - bd \end{pmatrix} \longleftrightarrow (ac - bd) + (ad + bc)i,$$

then

$$\begin{pmatrix} a & b \\ -b & a \end{pmatrix} \begin{pmatrix} c & d \\ -d & c \end{pmatrix} \longleftrightarrow (a + bi) \times (c + di),$$

which demonstrates that the correspondence is preserved under multiplication.

Example 2 Verify that the one-to-one correspondence between the matrices

$$\begin{pmatrix} x & y \\ -y & x \end{pmatrix}$$

and the complex numbers $x + yi$ is preserved under multiplication for the complex numbers $5 + i$ and $2 - 4i$.

Let

$$5 + i \longleftrightarrow \begin{pmatrix} 5 & 1 \\ -1 & 5 \end{pmatrix} \quad \text{and} \quad 2 - 4i \longleftrightarrow \begin{pmatrix} 2 & -4 \\ 4 & 2 \end{pmatrix}.$$

Then

$$(5 + i) \times (2 - 4i) = [(5)(2) - (1)(-4)] + [(5)(-4) + (1)(2)]i$$
$$= 14 - 18i,$$

and

$$\begin{pmatrix} 5 & 1 \\ -1 & 5 \end{pmatrix} \begin{pmatrix} 2 & -4 \\ 4 & 2 \end{pmatrix} = \begin{pmatrix} 14 & -18 \\ 18 & 14 \end{pmatrix}.$$

Since

$$14 - 18i \longleftrightarrow \begin{pmatrix} 14 & -18 \\ 18 & 14 \end{pmatrix},$$

then

$$(5 + i) \times (2 - 4i) \longleftrightarrow \begin{pmatrix} 5 & 1 \\ -1 & 5 \end{pmatrix} \begin{pmatrix} 2 & -4 \\ 4 & 2 \end{pmatrix}.$$

One reason why the algebra of matrices has become a valuable algebra for mathematicians to study is that many important mathematical systems are isomorphic to sets of matrices. Consider now one other such mathematical system which is of historical importance in the development of vector algebra.

A **quaternion** is an element of the form $a + bi + cj + dk$, where a, b, c, and d are real numbers. Two quaternions $a + bi + cj + dk$ and $a' + b'i + c'j + d'k$ are equal if and only if $a = a'$, $b = b'$, $c = c'$, and $d = d'$; two quaternions are added in the same manner as two linear functions in $i, j,$ and k. The product of any two quaternions may be obtained by assuming the distributive property of multiplication with respect to addition and defining $i^2 = j^2 = k^2 = -1$, $ij = -ji = k$, $jk = -kj = i$, and $ki = -ik = j$. Under these definitions of addition and multiplication of quaternions, it can be shown that the set of quaternions forms a ring. Furthermore, the ring of quaternions $a + bi + cj + dk$ can be shown to be isomorphic to two different sets of matrices: the set of square real matrices of order four of the form

$$\begin{pmatrix} a & b & c & d \\ -b & a & -d & c \\ -c & d & a & -b \\ -d & -c & b & a \end{pmatrix}$$

and the set of square complex matrices of order two of the form

$$\begin{pmatrix} a + bi & c + di \\ -c + di & a - bi \end{pmatrix}.$$

Example 3 Verify that the one-to-one correspondence between the matrices

$$\begin{pmatrix} a & b & c & d \\ -b & a & -d & c \\ -c & d & a & -b \\ -d & -c & b & a \end{pmatrix}$$

and the quaternions $a + bi + cj + dk$ is preserved under multiplication for the quaternions $2 + 3i + j + 4k$ and $1 - i + 3j - 2k$.

Let

$$2 + 3i + j + 4k \longleftrightarrow \begin{pmatrix} 2 & 3 & 1 & 4 \\ -3 & 2 & -4 & 1 \\ -1 & 4 & 2 & -3 \\ -4 & -1 & 3 & 2 \end{pmatrix}$$

and

$$1 - i + 3j - 2k \longleftrightarrow \begin{pmatrix} 1 & -1 & 3 & -2 \\ 1 & 1 & 2 & 3 \\ -3 & -2 & 1 & 1 \\ 2 & -3 & -1 & 1 \end{pmatrix}.$$

Then

$$(2 + 3i + j + 4k) \times (1 - i + 3j - 2k)$$
$$= [(2)(1) - (3)(-1) - (1)(3) - (4)(-2)]$$
$$+ [(2)(-1) + (3)(1) + (1)(-2) - (4)(3)]i$$
$$+ [(2)(3) - (3)(-2) + (1)(1) + (4)(-1)]j$$
$$+ [(2)(-2) + (3)(3) - (1)(-1) + (4)(1)]k$$
$$= 10 - 13i + 9j + 10k,$$

and

$$\begin{pmatrix} 2 & 3 & 1 & 4 \\ -3 & 2 & -4 & 1 \\ -1 & 4 & 2 & -3 \\ -4 & -1 & 3 & 2 \end{pmatrix} \begin{pmatrix} 1 & -1 & 3 & -2 \\ 1 & 1 & 2 & 3 \\ -3 & -2 & 1 & 1 \\ 2 & -3 & -1 & 1 \end{pmatrix} = \begin{pmatrix} 10 & -13 & 9 & 10 \\ 13 & 10 & -10 & 9 \\ -9 & 10 & 10 & 13 \\ -10 & -9 & -13 & 10 \end{pmatrix}.$$

Since

$$10 - 13i + 9j + 10k \longleftrightarrow \begin{pmatrix} 10 & -13 & 9 & 10 \\ 13 & 10 & -10 & 9 \\ -9 & 10 & 10 & 13 \\ -10 & -9 & -13 & 10 \end{pmatrix},$$

then the one-to-one correspondence is preserved under multiplication.

Exercises

Determine which of the following sets of elements form rings considering the operations as ordinary addition and multiplication unless otherwise stated.

1. The set of natural numbers.

2. The set of rational numbers.

3. The set of even integers.

4. The set of matrices

$$\begin{pmatrix} 1 & 0 \\ 0 & 1 \end{pmatrix}, \quad \begin{pmatrix} -1 & 0 \\ 0 & -1 \end{pmatrix}, \quad \text{and} \quad \begin{pmatrix} 0 & 0 \\ 0 & 0 \end{pmatrix}$$

under matrix addition and multiplication.

5. The set of square real matrices of order two of the form

$$\begin{pmatrix} a & b \\ 0 & 0 \end{pmatrix}$$

under matrix addition and multiplication.

In Exercises 6 through 9 verify that the one-to-one correspondence between the matrices

$$\begin{pmatrix} x & y \\ -y & x \end{pmatrix}$$

and the complex numbers $x + yi$ is preserved under the indicated operation for the given complex numbers.

6. $(5 + 2i) + (5 - 2i)$.   **7.** $(5 - 3i) + (2 - 4i)$.

8. $(3 + 4i) \times (2 - i)$.   **9.** $(i) \times (i)$.

10. Show that the reciprocal of the complex number $a + bi$ corresponds to the inverse of the matrix

$$\begin{pmatrix} a & b \\ -b & a \end{pmatrix}.$$

Consider an operation $\odot$ on the set S of nonsingular matrices of order two such that if $A, B \in S$, $A \odot B = AB - BA$. In Exercises 11 and 12 prove the indicated properties of $\odot$.

11. $A \odot B = -B \odot A$.   **12.** $A \odot (B \odot C) \neq (A \odot B) \odot C$.

In Exercises 13 and 14 verify that the one-to-one correspondence between the complex matrices

$$\begin{pmatrix} a + bi & c + di \\ -c + di & a - bi \end{pmatrix}$$

and the quaternions $a + bi + cj + dk$ is preserved under multiplication for the given quaternions.

13. $5i + j + k$ and $3 + i + 2j + 4k$.   **14.** $i + k$ and $3 + j$.

2-4 Rank of a Matrix

Before the discussion concerning the concept of the rank of a matrix, a brief review of the concept of linear dependence is in order.

Consider a set of functions, vectors, or elements $e_1, e_2, \ldots, e_n$. These functions, vectors, or elements $e_1, e_2, \ldots, e_n$ are said to be **linearly dependent** if there exists a set of scalars $k_1, k_2, \ldots, k_n$, not all zero, such that $k_1 e_1 + k_2 e_2 + \cdots + k_n e_n = 0$. If they are not linearly dependent, then they are said to be **linearly independent**.

Example 1 Show that the functions $f(x, y, z)$, $g(x, y, z)$, and $h(x, y, z)$ are linearly dependent where $f(x, y, z) = x - y + z$, $g(x, y, z) = x + y + 2z$, and $h(x, y, z) = 3x + y + 5z$.

If $f(x, y, z)$, $g(x, y, z)$, and $h(x, y, z)$ are linearly dependent functions, then there exists a set of scalars k_1, k_2, and k_3, not all zero, such that $k_1 f(x, y, z) + k_2 g(x, y, z) + k_3 h(x, y, z) = 0$. Consider $k_1(x - y + z) + k_2(x + y + 2z) + k_3(3x + y + 5z) = 0$. Then $(k_1 + k_2 + 3k_3)x + (-k_1 + k_2 + k_3)y + (k_1 + 2k_2 + 5k_3)z = 0$, and

$$\begin{cases} k_1 + k_2 + 3k_3 = 0 \\ -k_1 + k_2 + k_3 = 0 \\ k_1 + 2k_2 + 5k_3 = 0. \end{cases}$$

Solve simultaneously by adding the second equation to the first and third equations and obtain

$$\begin{cases} 2k_2 + 4k_3 = 0 \\ 3k_2 + 6k_3 = 0; \end{cases}$$

$k_2 = -2k_3$, and $k_1 = -k_3$. Let $k_3 = -1$. Then $k_2 = 2$, $k_1 = 1$, and $1(x - y + z) + 2(x + y + 2z) - 1(3x + y + 5z) = 0$. Hence, $f(x, y, z)$, $g(x, y, z)$, and $h(x, y, z)$ are linearly dependent functions.

Given any matrix A of order m by n, **square submatrices** of order r may be obtained by selecting the elements in any r rows and r columns of the matrix. The determinants of these square submatrices of order r are called **r-rowed minors** of the matrix A. The order of the largest square submatrix whose determinant has a nonzero value is called the **rank** of the matrix. The rank of any matrix other than a zero matrix cannot be zero, that of a matrix of order m by n cannot exceed either m or n, and that of a zero matrix is defined to be zero. In general, the rank of a matrix is equal to the largest number of linearly independent row vectors and column vectors since the value of the determinant of a square matrix is zero if the row vectors or column vectors are linearly dependent.

Example 2 Determine the rank of matrix A where

$$A = \begin{pmatrix} 4 & 2 & -1 & 3 \\ 0 & 5 & -1 & 2 \\ 12 & -4 & -1 & 5 \end{pmatrix}.$$

Since matrix A is of order 3 by 4, the rank of A cannot exceed three. Each of the 3-rowed minors of A, however, has value zero; that is,

$$\begin{vmatrix} 4 & 2 & -1 \\ 0 & 5 & -1 \\ 12 & -4 & -1 \end{vmatrix} = \begin{vmatrix} 4 & 2 & 3 \\ 0 & 5 & 2 \\ 12 & -4 & 5 \end{vmatrix} = \begin{vmatrix} 4 & -1 & 3 \\ 0 & -1 & 2 \\ 12 & -1 & 5 \end{vmatrix} = \begin{vmatrix} 2 & -1 & 3 \\ 5 & -1 & 2 \\ -4 & -1 & 5 \end{vmatrix} = 0.$$

Therefore, the rank of A cannot be three. Since the 2-rowed minor

$$\begin{vmatrix} 4 & 2 \\ 0 & 5 \end{vmatrix}$$

does not have value zero, the rank of A is two.

The difficulty in determining the rank of a matrix of high order may be reduced by considering three **elementary row transformations** on the matrix:

(i) interchanging the corresponding elements of two rows;

(ii) multiplication of the elements of any row by a nonzero scalar;

(iii) addition of a nonzero scalar multiple of the elements of any row to the corresponding elements of another row.

From Theorems 2-2, 2-4, and 2-5, it is evident that the rank of a matrix is not changed by an application of any of the elementary row transformations since under (i) only the signs of the values of some r-rowed minors are changed; under (ii) the values of some r-rowed minors are multiplied by a nonzero scalar; and under (iii) the value of every r-rowed minor is unchanged. These elementary row transformations may be applied in any sequence until the first nonzero element that appears in each row is equal to 1 and is positioned at the right of the first nonzero element of the preceding row. The matrix is then said to be in **echelon form**, and its rank may be easily determined.

Example 3 Determine the rank of matrix A where

$$A = \begin{pmatrix} 0 & 2 & 4 & 6 \\ 3 & -1 & 4 & -2 \\ 6 & -1 & 10 & -1 \end{pmatrix}.$$

Interchange the corresponding elements of rows one and two:

$$\begin{pmatrix} 3 & -1 & 4 & -2 \\ 0 & 2 & 4 & 6 \\ 6 & -1 & 10 & -1 \end{pmatrix};$$

multiply the elements of row one by (-2) and add the products to the corresponding elements of row three:

$$\begin{pmatrix} 3 & -1 & 4 & -2 \\ 0 & 2 & 4 & 6 \\ 0 & 1 & 2 & 3 \end{pmatrix};$$

multiply the elements of row two by $(-\frac{1}{2})$ and add the products to the corresponding elements of row three:

$$\begin{pmatrix} 3 & -1 & 4 & -2 \\ 0 & 2 & 4 & 6 \\ 0 & 0 & 0 & 0 \end{pmatrix};$$

multiply the elements of rows one and two by $(\frac{1}{3})$ and $(\frac{1}{2})$, respectively:

$$\begin{pmatrix} 1 & -\frac{1}{3} & \frac{4}{3} & -\frac{2}{3} \\ 0 & 1 & 2 & 3 \\ 0 & 0 & 0 & 0 \end{pmatrix}.$$

The matrix A is now in echelon form, and it is evident immediately that every 3-rowed minor has value zero. The rank of A is two since

$$\begin{vmatrix} 1 & -\frac{1}{3} \\ 0 & 1 \end{vmatrix} \neq 0.$$

Each of the elementary row transformations on a matrix may be viewed as a premultiplication of the matrix by a conformable matrix obtained by performing that elementary row transformation on the identity matrix. For example, if the matrix whose rank is to be determined is of order 3 by n, then the first elementary row transformation may be accomplished by premultiplying the given matrix by one of the matrices

$$\begin{pmatrix} 0 & 0 & 1 \\ 0 & 1 & 0 \\ 1 & 0 & 0 \end{pmatrix}, \quad \begin{pmatrix} 0 & 1 & 0 \\ 1 & 0 & 0 \\ 0 & 0 & 1 \end{pmatrix}, \quad \text{or} \quad \begin{pmatrix} 1 & 0 & 0 \\ 0 & 0 & 1 \\ 0 & 1 & 0 \end{pmatrix}. \tag{2-14}$$

The second elementary row transformation also may be accomplished by premultiplying the given matrix by one of the matrices

$$\begin{pmatrix} k & 0 & 0 \\ 0 & 1 & 0 \\ 0 & 0 & 1 \end{pmatrix}, \quad \begin{pmatrix} 1 & 0 & 0 \\ 0 & k & 0 \\ 0 & 0 & 1 \end{pmatrix}, \quad \text{or} \quad \begin{pmatrix} 1 & 0 & 0 \\ 0 & 1 & 0 \\ 0 & 0 & k \end{pmatrix}; \tag{2-15}$$

in addition, the third elementary row transformation may be accomplished by premultiplying the given matrix by one of the matrices

$$\begin{pmatrix} 1 & k & 0 \\ 0 & 1 & 0 \\ 0 & 0 & 1 \end{pmatrix}, \quad \begin{pmatrix} 1 & 0 & k \\ 0 & 1 & 0 \\ 0 & 0 & 1 \end{pmatrix}, \quad \begin{pmatrix} 1 & 0 & 0 \\ k & 1 & 0 \\ 0 & 0 & 1 \end{pmatrix},$$

$$\begin{pmatrix} 1 & 0 & 0 \\ 0 & 1 & k \\ 0 & 0 & 1 \end{pmatrix}, \quad \begin{pmatrix} 1 & 0 & 0 \\ 0 & 1 & 0 \\ k & 0 & 1 \end{pmatrix}, \quad \text{or} \quad \begin{pmatrix} 1 & 0 & 0 \\ 0 & 1 & 0 \\ 0 & k & 1 \end{pmatrix}. \tag{2-16}$$

The matrices which permit the performance of elementary row transformations are called **elementary row transformation matrices**. For example, the matrices of (2-14), (2-15), and (2-16) are elementary row transformation matrices of order three. Note that every elementary row transformation matrix is nonsingular.

It is possible to show that every nonsingular matrix is a product of elementary row transformation matrices. Then every nonsingular matrix A has a nonsingular inverse A^{-1} which is equal to a product of elementary row transformation matrices. The product of elementary row transformation matrices transforming A into I is A^{-1}. Note carefully that the elementary row transformation matrices used must be multiplied in the reverse order of their application.

Example 4 Determine the inverse of A where

$$A = \begin{pmatrix} 3 & 1 \\ 4 & 2 \end{pmatrix}.$$

Since A is nonsingular, perform the elementary row transformations on A until the identity matrix I is obtained. The product of the matrices representing the transformations will be equal to A^{-1}. For example, multiply the elements of row one by $(-\frac{4}{3})$ and add the products to the corresponding elements of row two:

$$\begin{pmatrix} 1 & 0 \\ -\frac{4}{3} & 1 \end{pmatrix}\begin{pmatrix} 3 & 1 \\ 4 & 2 \end{pmatrix} = \begin{pmatrix} 3 & 1 \\ 0 & \frac{2}{3} \end{pmatrix};$$

multiply the elements of row one by $(\frac{1}{3})$, then multiply the elements of row two by $(\frac{3}{2})$:

$$\begin{pmatrix} 1 & 0 \\ 0 & \frac{3}{2} \end{pmatrix}\begin{pmatrix} \frac{1}{3} & 0 \\ 0 & 1 \end{pmatrix}\begin{pmatrix} 3 & 1 \\ 0 & \frac{2}{3} \end{pmatrix} = \begin{pmatrix} 1 & \frac{1}{3} \\ 0 & 1 \end{pmatrix};$$

multiply the elements of row two by $(-\frac{1}{3})$ and add the products to the corresponding elements of row one:

$$\begin{pmatrix} 1 & -\frac{1}{3} \\ 0 & 1 \end{pmatrix}\begin{pmatrix} 1 & \frac{1}{3} \\ 0 & 1 \end{pmatrix} = \begin{pmatrix} 1 & 0 \\ 0 & 1 \end{pmatrix}.$$

Hence,

$$A^{-1} = \begin{pmatrix} 1 & -\frac{1}{3} \\ 0 & 1 \end{pmatrix}\begin{pmatrix} 1 & 0 \\ 0 & \frac{3}{2} \end{pmatrix}\begin{pmatrix} \frac{1}{3} & 0 \\ 0 & 1 \end{pmatrix}\begin{pmatrix} 1 & 0 \\ -\frac{4}{3} & 1 \end{pmatrix} = \begin{pmatrix} 1 & -\frac{1}{2} \\ -2 & \frac{3}{2} \end{pmatrix}.$$

Exercises

1. Show that the functions $2x + z$, $x + y$, and $2y - z$ are linearly dependent.

2. Show that the functions $x + y - z$, $2x + y + 3z$, and $x + 2y + 4z$ are linearly independent.

In Exercises 3 through 5 determine the rank of the given matrix.

3. $\begin{pmatrix} 4 & 2 & 3 \\ 8 & 5 & 2 \\ 12 & -4 & 5 \end{pmatrix}.$
   4. $\begin{pmatrix} 2 & 3 & 3 \\ 3 & 6 & 12 \\ 2 & 4 & 8 \end{pmatrix}.$
   5. $\begin{pmatrix} 3 & 4 & 0 & 2 \\ 6 & 8 & 0 & 4 \\ 1 & 0 & 1 & 1 \end{pmatrix}.$

6. Prove that the rank of any skew-symmetric matrix cannot be one.

7. Determine the maximum number of elementary row transformations necessary to transform a square matrix of order n to echelon form.

8. State the general forms of the elementary row transformation matrices of order two.

In Exercises 9 through 11 express the inverse of the given matrix as a product of elementary row transformation matrices.

9. $\begin{pmatrix} 6 & 2 \\ 2 & 1 \end{pmatrix}.$
   10. $\begin{pmatrix} 1 & 2 \\ -2 & 3 \end{pmatrix}.$
   11. $\begin{pmatrix} 1 & -2 & 0 \\ -2 & 3 & 1 \\ -5 & 9 & 3 \end{pmatrix}.$

12. Express the matrix given in Exercise 9 as a product of elementary row transformation matrices.

2-5 Systems of Linear Equations

Consider a system of m linear equations in n variables $x_1, x_2, \ldots, x_n$:

$$\begin{cases} a_{11}x_1 + a_{12}x_2 + \cdots + a_{1n}x_n = c_1 \\ a_{21}x_1 + a_{22}x_2 + \cdots + a_{2n}x_n = c_2 \\ \ldots \\ a_{m1}x_1 + a_{m2}x_2 + \cdots + a_{mn}x_n = c_m. \end{cases} \qquad (2\text{-}17)$$

The system of (2-17) may be written in matrix form as

$$\begin{pmatrix} a_{11} & a_{12} & \cdots & a_{1n} \\ a_{21} & a_{22} & \cdots & a_{2n} \\ \cdots & \cdots & \cdots & \cdots \\ a_{m1} & a_{m2} & \cdots & a_{mn} \end{pmatrix} \begin{pmatrix} x_1 \\ x_2 \\ \cdots \\ x_n \end{pmatrix} = \begin{pmatrix} c_1 \\ c_2 \\ \cdots \\ c_m \end{pmatrix}. \qquad (2\text{-}18)$$

The matrix $((a_{ij}))$ of order m by n is called the **matrix of coefficients**. The matrix composed of the mn elements a_{ij} plus an additional column whose elements are the constants c_i is called the **augmented matrix** of the system; that is,

$$\begin{pmatrix} a_{11} & a_{12} & \cdots & a_{1n} & c_1 \\ a_{21} & a_{22} & \cdots & a_{2n} & c_2 \\ \cdots & \cdots & \cdots & \cdots & \cdots \\ a_{m1} & a_{m2} & \cdots & a_{mn} & c_m \end{pmatrix} \qquad (2\text{-}19)$$

is the augmented matrix of (2-17). Note that the rank of the augmented matrix is either equal to or one greater than the rank of the matrix of coefficients.

A solution, if it exists, to a system of linear equations (2-17) may be determined by transforming the augmented matrix using the elementary row transformations until the matrix is in echelon form. The system of linear equations represented by the transformed augmented matrix will be an equivalent system to (2-17); "equivalent" is used in the sense that the solution set is not changed.

If the system of linear equations (2-17) is such that the equations are all satisfied simultaneously by at least one set of values of the variables, then it is said to be **consistent**. The system is said to be **inconsistent** if the equations are not satisfied simultaneously by any set of values of the variables.

Example 1 Solve the system of linear equations by performing the elementary row transformations on the augmented matrix:

$$\begin{cases} 2x - y + z = 1 \\ x + y - z = 2 \\ 3x - y + z = 0. \end{cases}$$

The augmented matrix is

$$\begin{pmatrix} 2 & -1 & 1 & 1 \\ 1 & 1 & -1 & 2 \\ 3 & -1 & 1 & 0 \end{pmatrix}.$$

Interchange the corresponding elements of rows one and two:

$$\begin{pmatrix} 1 & 1 & -1 & 2 \\ 2 & -1 & 1 & 1 \\ 3 & -1 & 1 & 0 \end{pmatrix};$$

multiply the elements of row one by (-2) and add the products to the corresponding elements of row two:

$$\begin{pmatrix} 1 & 1 & -1 & 2 \\ 0 & -3 & 3 & -3 \\ 3 & -1 & 1 & 0 \end{pmatrix};$$

multiply the elements of row one by (-3) and add the products to the corresponding elements of row three:

$$\begin{pmatrix} 1 & 1 & -1 & 2 \\ 0 & -3 & 3 & -3 \\ 0 & -4 & 4 & -6 \end{pmatrix};$$

multiply the elements of row two by $(-\frac{1}{3})$:

$$\begin{pmatrix} 1 & 1 & -1 & 2 \\ 0 & 1 & -1 & 1 \\ 0 & -4 & 4 & -6 \end{pmatrix};$$

multiply the elements of row two by (4) and add the products to the corresponding elements of row three:

$$\begin{pmatrix} 1 & 1 & -1 & 2 \\ 0 & 1 & -1 & 1 \\ 0 & 0 & 0 & -2 \end{pmatrix};$$

multiply the elements of row three by $(-\frac{1}{2})$:

$$\begin{pmatrix} 1 & 1 & -1 & 2 \\ 0 & 1 & -1 & 1 \\ 0 & 0 & 0 & 1 \end{pmatrix}.$$

The transformed augmented matrix is now in echelon form and represents the system of linear equations

$$\begin{cases} x + y - z = 2 \\ \quad\quad y - z = 1 \\ \quad\quad\quad\quad 0 = 1. \end{cases}$$

Since no values of x, y, and z exist such that 0 equals 1, the system is inconsistent.

Note that in Example 1 the rank of the augmented matrix is three while the rank of the matrix of coefficients is two.

Example 2 Solve the system of linear equations by performing the elementary row transformations on the augmented matrix:

$$\begin{cases} x + 3y + z = 6 \\ 3x - 2y - 8z = 7 \\ 4x + 5y - 3z = 17. \end{cases}$$

The augmented matrix is

$$\begin{pmatrix} 1 & 3 & 1 & 6 \\ 3 & -2 & -8 & 7 \\ 4 & 5 & -3 & 17 \end{pmatrix}.$$

Multiply the elements of row one by (-3) and add the products to the corresponding elements of row two:

$$\begin{pmatrix} 1 & 3 & 1 & 6 \\ 0 & -11 & -11 & -11 \\ 4 & 5 & -3 & 17 \end{pmatrix};$$

multiply the elements of row one by (-4) and add the products to the corresponding elements of row three:

$$\begin{pmatrix} 1 & 3 & 1 & 6 \\ 0 & -11 & -11 & -11 \\ 0 & -7 & -7 & -7 \end{pmatrix};$$

multiply the elements of row two by $(-\frac{1}{11})$:

$$\begin{pmatrix} 1 & 3 & 1 & 6 \\ 0 & 1 & 1 & 1 \\ 0 & -7 & -7 & -7 \end{pmatrix};$$

multiply the elements of row two by (7) and add the products to the corresponding elements of row three:

$$\begin{pmatrix} 1 & 3 & 1 & 6 \\ 0 & 1 & 1 & 1 \\ 0 & 0 & 0 & 0 \end{pmatrix}.$$

The transformed augmented matrix is now in echelon form and represents the system of linear equations

$$\begin{cases} x + 3y + z = 6 \\ y + z = 1 \\ 0 = 0. \end{cases}$$

Hence, there exists an infinite number of solutions of the form $x = 3 + 2z$ and $y = 1 - z$. The value of the variable z may be assigned arbitrarily, thus fixing the values of x and y.

Note that in Example 2 the rank of both the augmented matrix and the matrix of coefficients is two. In general, a necessary and sufficient condition for a system of linear equations to be consistent is that the augmented matrix and the matrix of coefficients be of the same rank.

For the reader's information, some general statements about systems of m linear equations in n variables follow. Although the statements may be proved, they will be assumed throughout this book. If the system (2-17) is consistent in the case of $m > n$ (more equations than variables), the equations are necessarily linearly dependent. Furthermore, if the rank of the augmented matrix is n, a unique solution exists; however, if the rank r of the augmented matrix is less than n, a solution may be obtained by choosing arbitrary values for $n - r$ of the variables and proceeding to determine values of the remaining variables; that is, an infinite number of solutions exist. Note that when $m < n$, values for at least $n - m$ variables must be chosen arbitrarily if the system is consistent.

A system of linear homogeneous equations is of the form

$$\begin{cases} a_{11}x_1 + a_{12}x_2 + \cdots + a_{1n}x_n = 0 \\ a_{21}x_1 + a_{22}x_2 + \cdots + a_{2n}x_n = 0 \\ \cdots \\ a_{m1}x_1 + a_{m2}x_2 + \cdots + a_{mn}x_n = 0. \end{cases} \tag{2-20}$$

The system (2-20) may be expressed in matrix form as $AX = 0$, where A is the matrix of coefficients, X is the column matrix of variables $x_1, x_2, \ldots, x_n$, and 0 is the zero column matrix of order m by 1. Such a system necessarily is consistent since it has the *trivial* solution $x_1 = x_2 = \cdots = x_n = 0$. If a *nontrivial* solution exists, an infinite number of solutions exists; that is, if $t_1, t_2, \ldots, t_n$ represents a nontrivial solution, then $kt_1, kt_2, \ldots, kt_n$ for any real number k is also a solution. This property of a system of linear homogeneous equations may be proved by direct substitution in (2-20).

Exercises

In Exercises 1 and 2 solve the system of linear equations by performing the elementary row transformations on the augmented matrix.

1. $\begin{cases} 2x - 4y + 5z = 10 \\ 2x - 11y + 10z = 36 \\ 4x - y + 5z = -6. \end{cases}$ **2.** $\begin{cases} x + 2y - z = 2 \\ x - 2y - z = 5 \\ 2x + 4y - 2z = 7. \end{cases}$

3. Determine k such that the system of linear equations is consistent:

$$\begin{cases} 2x + y - z = 12 \\ x - y - 2z = -3 \\ 3y + 3z = k. \end{cases}$$

4. Prove that a system of n linear homogeneous equations in n variables has a nontrivial solution if and only if the rank of the matrix of coefficients is less than n.

In Exercises 5 and 6 determine whether or not the system of linear homogeneous equations has nontrivial solutions. (See Exercise 4.)

5. $\begin{cases} 3x + 8y + 2z = 0 \\ 2x + y + 3z = 0 \\ -5x - y + z = 0. \end{cases}$ **6.** $\begin{cases} x + 3y - 5z = 0 \\ 3x - y + 5z = 0 \\ 3x + 2y - z = 0. \end{cases}$

chapter 3

Transformations
of the Plane

3-1 Mappings

In elementary mathematics a function may be considered as a rule which associates with each element of some set A an element of some set B. Except in advanced mathematics, the sets A and B are usually subsets of the set of real numbers. For example, if A consists of the set of real numbers x, and B consists of the set of nonnegative real numbers y, then $y = x^2$ represents a function; that is, to each $x \in A$ the rule $y = x^2$ associates an element $y \in B$ where y is the square of x. In this section some additional terminology for the concept of a function is introduced. This terminology is based on geometric concepts and lends itself to the description of the problems to be discussed in this chapter.

A **single-valued mapping** T of a set A *into* a set B is a rule which associates

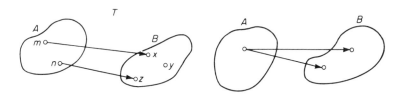

Figure 3-1   Figure 3-2

with each element $a \in A$ a unique element $b \in B$. The element b is called the **image** of a under the mapping T and is denoted by $b = T(a)$. For example, consider two sets, $A = \{m, n\}$ and $B = \{x, y, z\}$. Let T be a rule which associates x and z with m and n, respectively; then T is a single-valued mapping of A into B such that $x = T(m)$ and $z = T(n)$. This mapping is illustrated geometrically by Figure 3-1.

Figure 3-2 illustrates a situation which cannot be described as a single-valued mapping of A into B since no element of set A may have more than one image.

If T is a single-valued mapping of A into B such that each element of B is the image of some element of A, then T is a single-valued mapping of A **onto** B. In Figure 3-1, T is a mapping of A into B, but is not a mapping of A onto B. Consider two sets, $A = \{1, 2, 3\}$ and $B = \{a, b, c\}$. Let T be a rule which associates a, b, and c with 2, 1, and 3, respectively (Figure 3-3); then T is a single-valued mapping of A onto B. Every mapping of A onto B is necessarily a mapping of A into B; however, the converse is not true.

As another example of a single-valued mapping of A onto B, let $A = \{1, 2, 3\}$, $B = \{m, n\}$, and T be a rule such that $T(1) = m$, $T(2) = m$, and $T(3) = n$ (Figure 3-4). Notice that while each element of B is the image of some element of A, m is the image of both 1 and 2. The mapping T is nevertheless a single-valued mapping of A onto B. A distinction exists, however, between the two single-valued mappings illustrated in Figures 3-3

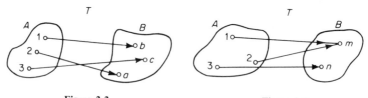

Figure 3-3   Figure 3-4

and 3-4. The mapping in Figure 3-3 is called a **one-to-one mapping** of A onto B and is a special type of onto mapping. Every one-to-one mapping of A onto B is a single-valued mapping of A onto B, but not every single-valued mapping of A onto B is a one-to-one mapping of A onto B. The mapping in Figure 3-4 is not a one-to-one mapping of A onto B.

If T is a one-to-one mapping of A onto B, then each element of B is the image of exactly one element of A. Hence, it is possible to define an inverse mapping of B onto A, denoted by T^{-1}. The **inverse mapping** T^{-1} is such that it associates with each element $b \in B$ the element $a \in A$ that has b as its image under T; that is, $T^{-1}(T(a)) = a$ for all $a \in A$.

Example 1 Let N be the set of integers and $n \in N$. Let T be a single-valued mapping of N into N. Determine if T is a mapping of N onto N where
(a) $T(n) = n + 1$; **(b)** $T(n) = 2n + 1$.

(a) T is a mapping of N onto N since each integer is the unique image of another integer one unit less. Furthermore, T is one-to-one mapping of N onto N since each integer is the image of one and only one other integer.

(b) T is not a mapping of N onto N since no even integer is the image of another integer under the mapping; that is, $2n + 1$ cannot be even for any integer n.

Example 2 Let T be a single-valued mapping of A into B where $A = \{a, b\}$ and $B = \{x, y\}$. Define the possible mappings.

There are four possible mappings T of A into B:

(i) $T(a) = x$ and $T(b) = x$;   (ii) $T(a) = x$ and $T(b) = y$;

(iii) $T(a) = y$ and $T(b) = x$;   (iv) $T(a) = y$ and $T(b) = y$.

Note that the mappings T described in (ii) and (iii) are one-to-one mappings of A onto B.

Some important mappings of the set of points on a plane into itself are illustrated in the remaining sections of this chapter.

Exercises

1. Let R be the set of real numbers and $x \in R$. Let T be a single-valued mapping of R into R. Determine if T is a mapping of R onto R where
   (a) $T(x) = x + 3$;   (b) $T(x) = 2$;
   (c) $T(x) = x^3$;   (d) $T(x) = x^3 - x$.

2. Let A and B be sets with the same finite number of elements. Prove that T is a one-to-one mapping of A onto B if T is a single-valued mapping of A onto B.

3. Determine the possible number of mappings T in Exercise 2 if A and B are sets with n elements.

4. Determine the possible number of single-valued mappings T of a set A into itself if A is a set with n elements.

5. Let $A = \{1, 2, 3\}$ and $B = \{a, b\}$. Let T be a single-valued mapping of A into B such that $T(1) = b$, $T(2) = a$, and $T(3) = b$. Determine the single-valued mapping T^{-1}, if it exists.

3-2 Rotations

It is sometimes necessary or desirable to consider a mapping of the set of points on a plane into itself in order to simplify a problem under examination. Such a mapping is called a **point transformation** or a **transformation of the plane**; that is, a transformation of the plane is a rule by which each point P on the plane is transformed or *mapped* onto a point P' on the plane.

A transformation of the plane may be considered as a movement of the points on the plane to the position of their respective image points. Each transformation of the plane may be described algebraically by means of a set of equations relating the rectangular cartesian coordinates of each point on the plane to the coordinates of its image point. There exist various types of transformations of the plane such as rotations, reflections, dilations, magnifications, shears, projections, and inversions, among others. In this section and in subsequent sections some of the most important transformations are considered and represented in matrix form.

Consider a **rotation of the plane** about the origin through an angle θ, retaining the same rectangular cartesian coordinate reference system. Let $P: (x, y)$ be any point on the plane. Under a rotation of the plane about the

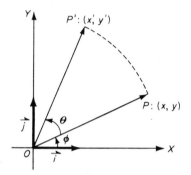

Figure 3-5

origin, each point P is mapped onto a point $P': (x', y')$ as in Figure 3-5. Note that $|\overrightarrow{OP}| = |\overrightarrow{OP'}|$. Then

$$\overrightarrow{OP} = x\vec{i} + y\vec{j}$$
$$= |\overrightarrow{OP}| \cos \phi \vec{i} + |\overrightarrow{OP}| \sin \phi \vec{j}$$
$$= |\overrightarrow{OP'}| \cos \phi \vec{i} + |\overrightarrow{OP'}| \sin \phi \vec{j},$$

and

$$\overrightarrow{OP'} = x'\vec{i} + y'\vec{j}$$
$$= |\overrightarrow{OP'}| \cos (\theta + \phi)\vec{i} + |\overrightarrow{OP'}| \sin (\theta + \phi)\vec{j}$$
$$= |\overrightarrow{OP'}| (\cos \theta \cos \phi - \sin \theta \sin \phi)\vec{i}$$
$$+ |\overrightarrow{OP'}| (\sin \theta \cos \phi + \cos \theta \sin \phi)\vec{j}$$
$$= (x \cos \theta - y \sin \theta)\vec{i} + (x \sin \theta + y \cos \theta)\vec{j}.$$

Hence,

$$\begin{cases} x' = x \cos \theta - y \sin \theta \\ y' = x \sin \theta + y \cos \theta. \end{cases} \qquad (3\text{-}1)$$

The relationship between the coordinates of each point P and its image point P' may be expressed in matrix form as

$$\begin{pmatrix} x' \\ y' \end{pmatrix} = \begin{pmatrix} \cos \theta & -\sin \theta \\ \sin \theta & \cos \theta \end{pmatrix} \begin{pmatrix} x \\ y \end{pmatrix}. \qquad (3\text{-}2)$$

The matrix

$$\begin{pmatrix} \cos \theta & -\sin \theta \\ \sin \theta & \cos \theta \end{pmatrix} \qquad (3\text{-}3)$$

of the rotation transformation defined by (3-2) is called a **rotation matrix**.

The rotation matrix (3-3) may be considered as an *operator* that maps each point (x, y) on the plane onto its image point

$$(x \cos \theta - y \sin \theta, \ x \sin \theta + y \cos \theta)$$

when the plane is rotated about the origin through an angle θ. When considered as an operator, a matrix of the form (3-3) may be used to represent, describe, or characterize a rotation of the plane about the origin through an angle θ. Note that the determinant of every rotation matrix is equal to one. Furthermore, either the row or column vectors of a rotation matrix may be considered to be a pair of orthogonal unit vectors.

Example 1 Determine the coordinates of the image point of P: $(5, \sqrt{3})$ under a rotation of the plane about the origin through an angle of 30° (Figure 3-6).

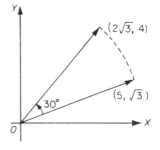

Since the angle of rotation θ is 30°, $\cos \theta = \sqrt{3}/2$ and $\sin \theta = \frac{1}{2}$. Therefore, by (3-3), the rotation matrix for the transformation of the plane is

$$\begin{pmatrix} \frac{\sqrt{3}}{2} & -\frac{1}{2} \\ \frac{1}{2} & \frac{\sqrt{3}}{2} \end{pmatrix}.$$

Hence, by (3-2), the image point (x', y') of P: $(5, \sqrt{3})$ is given as

Figure 3-6

$$\begin{pmatrix} x' \\ y' \end{pmatrix} = \begin{pmatrix} \frac{\sqrt{3}}{2} & -\frac{1}{2} \\ \frac{1}{2} & \frac{\sqrt{3}}{2} \end{pmatrix} \begin{pmatrix} 5 \\ \sqrt{3} \end{pmatrix} = \begin{pmatrix} 2\sqrt{3} \\ 4 \end{pmatrix};$$

that is, the coordinates of the image point of P: $(5, \sqrt{3})$ under a rotation of the plane about the origin through an angle of 30° are $(2\sqrt{3}, 4)$.

Example 2 Determine the equation satisfied by the set of image points of the locus of $x^2 - 2xy + y^2 - \sqrt{2}\,x - \sqrt{2}\,y = 0$ under a rotation of the plane about the origin through an angle of 45° (Figure 3-7).

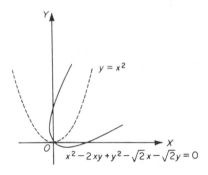

Figure 3-7

Since the angle of rotation θ is $45°$, $\cos\theta = \sqrt{2}/2$ and $\sin\theta = \sqrt{2}/2$. The image point (x', y') of each point (x, y) is given by the matrix equation (3-8) as

$$\begin{pmatrix} x' \\ y' \end{pmatrix} = \begin{pmatrix} \frac{\sqrt{2}}{2} & -\frac{\sqrt{2}}{2} \\ \frac{\sqrt{2}}{2} & \frac{\sqrt{2}}{2} \end{pmatrix}\begin{pmatrix} x \\ y \end{pmatrix}.$$

Then

$$\begin{pmatrix} x \\ y \end{pmatrix} = \begin{pmatrix} \frac{\sqrt{2}}{2} & \frac{\sqrt{2}}{2} \\ -\frac{\sqrt{2}}{2} & \frac{\sqrt{2}}{2} \end{pmatrix}\begin{pmatrix} x' \\ y' \end{pmatrix}.$$

Hence, replacing x by $\sqrt{2}\,x'/2 + \sqrt{2}\,y'/2$ and y by $-\sqrt{2}\,x'/2 + \sqrt{2}\,y'/2$ in $x^2 - 2xy + y^2 - \sqrt{2}\,x - \sqrt{2}\,y = 0$, the equation satisfied by the set of image points becomes

$$(\tfrac{\sqrt{2}}{2}x' + \tfrac{\sqrt{2}}{2}y')^2 - 2(\tfrac{\sqrt{2}}{2}x' + \tfrac{\sqrt{2}}{2}y')(-\tfrac{\sqrt{2}}{2}x' + \tfrac{\sqrt{2}}{2}y')$$
$$+ (-\tfrac{\sqrt{2}}{2}x' + \tfrac{\sqrt{2}}{2}y')^2 - \sqrt{2}(\tfrac{\sqrt{2}}{2}x' + \tfrac{\sqrt{2}}{2}y') - \sqrt{2}(-\tfrac{\sqrt{2}}{2}x' + \tfrac{\sqrt{2}}{2}y') = 0.$$

Simplifying, $y' = (x')^2$; that is, $y = x^2$ since (x', y') are the coordinates of each image point with reference to the xy-axes.

A point is called a **fixed point** under a transformation of the plane if it is mapped onto itself. Note that when $\theta \neq 360° \, k$ for any integer k, the origin is the only fixed point under a rotation of the plane about the origin through an angle θ. When $\theta = 360° \, k$ for some integer k, every point on the plane is a fixed point, and the rotation matrix (3-3) is equal to the identity matrix of order two. A rotation of the plane about the origin represented by an identity matrix is called the **identity transformation**.

Occasionally it is necessary to apply one transformation of the plane after another. If T_1 and T_2 are matrices representing transformations, then the **product transformation** T_1 followed by T_2 is denoted by T_2T_1 and defined such that

$$T_2T_1\begin{pmatrix} x \\ y \end{pmatrix} = T_2\left[T_1\begin{pmatrix} x \\ y \end{pmatrix}\right]. \tag{3-4}$$

A rotation of the plane about the origin through an angle θ is a one-to-one mapping of the set of points on the plane onto itself. Hence, an inverse transformation exists for each rotation. Consider a rotation of the plane about the origin through an angle $(-\theta)$. By (3-3), the rotation matrix of this transformation is

$$\begin{pmatrix} \cos(-\theta) & -\sin(-\theta) \\ \sin(-\theta) & \cos(-\theta) \end{pmatrix};$$

that is,

$$\begin{pmatrix} \cos\theta & \sin\theta \\ -\sin\theta & \cos\theta \end{pmatrix} \tag{3-5}$$

since $\cos(-\theta) = \cos\theta$ and $\sin(-\theta) = -\sin\theta$. Then, since

$$\begin{pmatrix} \cos\theta & \sin\theta \\ -\sin\theta & \cos\theta \end{pmatrix}\begin{pmatrix} \cos\theta & -\sin\theta \\ \sin\theta & \cos\theta \end{pmatrix} = \begin{pmatrix} 1 & 0 \\ 0 & 1 \end{pmatrix},$$

a rotation of the plane about the origin through an angle $(-\theta)$ is the inverse transformation of a rotation of the plane about the origin through an angle θ.

Exercises

In Exercises 1 through 6 determine the image point of the point P under a rotation of the plane about the origin through the angle θ.

1. P: $(2, -3)$; $\theta = 90°$.   2. P: $(-5, -2)$; $\theta = 180°$.

3. P: $(\sqrt{3}, 1)$; $\theta = 30°$.   4. P: $(\sqrt{3}, 1)$; $\theta = -60°$.

5. P: $(1, 2)$; $\theta = 45°$.   6. P: $(2, \sqrt{3})$; $\theta = 30°$.

7. Determine the equation satisfied by the set of image points of the locus of $x^2 - xy + y^2 = 16$ under a rotation of the plane about the origin through an angle of 45°.

8. Determine the equation satisfied by the set of image points of the locus of $16x^2 + 24xy + 9y^2 - 60x + 80y = 0$ under a rotation of the plane about the origin through an acute angle $\theta = \arctan\frac{4}{3}$.

9. Determine the equation satisfied by the set of image points of the locus of $x^2 + y^2 = r^2$ under a rotation of the plane about the origin through an angle θ.

10. Determine whether or not the matrix

$$\begin{pmatrix} -\frac{\sqrt{3}}{2} & \frac{1}{2} \\ \frac{1}{2} & \frac{\sqrt{3}}{2} \end{pmatrix}$$

may represent a rotation of the plane about the origin.

11. Prove that the multiplication of rotation matrices of the form (3-3) is **(a)** closed; **(b)** commutative.

12. Determine a rotation matrix that maps the points $(3, 4)$ and $(1, -2)$ onto the points $(-4, 3)$ and $(2, 1)$, respectively.

13. A geometric property that is unchanged under a given transformation is called an **invariant** under that transformation. Prove that the distance between two points on a plane is invariant under a rotation of the plane about the origin.

3-3 Reflections, Dilations, and Magnifications

In this section some additional types of transformations of the plane are presented.

Consider the transformations of the plane represented by the matrices

$$\begin{pmatrix} -1 & 0 \\ 0 & 1 \end{pmatrix} \quad \text{and} \quad \begin{pmatrix} 1 & 0 \\ 0 & -1 \end{pmatrix}. \tag{3-6}$$

These matrices map each point (x, y) on the plane onto the points $(-x, y)$ and $(x, -y)$, respectively, and represent one-to-one mappings of a plane onto itself that are called **reflections of the plane.** Each point on the plane is mapped onto its "mirror image" with respect to one of the coordinate axes by these **reflection matrices** (3-6). Note that under the reflection described by

$$\begin{pmatrix} -1 & 0 \\ 0 & 1 \end{pmatrix}$$

the points on the y-axis are fixed points; under the reflection described by

$$\begin{pmatrix} 1 & 0 \\ 0 & -1 \end{pmatrix}$$

the points on the x-axis are fixed points.

Another pair of matrices that represent reflections of the plane is given by

$$\begin{pmatrix} 0 & 1 \\ 1 & 0 \end{pmatrix} \quad \text{and} \quad \begin{pmatrix} 0 & -1 \\ -1 & 0 \end{pmatrix}. \tag{3-7}$$

The matrix

$$\begin{pmatrix} 0 & 1 \\ 1 & 0 \end{pmatrix}$$

maps each point (x, y) onto (y, x); that is, each point on the plane is mapped onto its mirror image with respect to the line $y = x$. The matrix

$$\begin{pmatrix} 0 & -1 \\ -1 & 0 \end{pmatrix}$$

maps each point (x, y) onto $(-y, -x)$; that is, each point on the plane is mapped onto its mirror image with respect to the line $y = -x$.

It is interesting and important to note that the product of any two of the reflection matrices of (3-6) and (3-7) is a rotation matrix. In general, the product of any two reflections of the plane with respect to intersecting lines passing through the origin is a rotation of the plane about the origin. Also note that the determinant of each of the reflection matrices in (3-6) and (3-7) is equal to -1.

The reflection matrices of (3-7) may be considered as products of the reflection matrices of (3-6) and rotation matrices of the form (3-3). For example,

$$\begin{pmatrix} 0 & 1 \\ 1 & 0 \end{pmatrix} = \begin{pmatrix} \frac{\sqrt{2}}{2} & -\frac{\sqrt{2}}{2} \\ \frac{\sqrt{2}}{2} & \frac{\sqrt{2}}{2} \end{pmatrix} \begin{pmatrix} 1 & 0 \\ 0 & -1 \end{pmatrix} \begin{pmatrix} \frac{\sqrt{2}}{2} & \frac{\sqrt{2}}{2} \\ -\frac{\sqrt{2}}{2} & \frac{\sqrt{2}}{2} \end{pmatrix}; \qquad (3\text{-}8)$$

$$\begin{pmatrix} 0 & -1 \\ -1 & 0 \end{pmatrix} = \begin{pmatrix} \frac{\sqrt{2}}{2} & \frac{\sqrt{2}}{2} \\ -\frac{\sqrt{2}}{2} & \frac{\sqrt{2}}{2} \end{pmatrix} \begin{pmatrix} 1 & 0 \\ 0 & -1 \end{pmatrix} \begin{pmatrix} \frac{\sqrt{2}}{2} & -\frac{\sqrt{2}}{2} \\ \frac{\sqrt{2}}{2} & \frac{\sqrt{2}}{2} \end{pmatrix}. \qquad (3\text{-}9)$$

Reflection matrices other than those of (3-6) and (3-7) exist. These reflection matrices, which map each point on the plane onto its mirror image with respect to a line l, can be shown to be equal to

$$T^{-1}RT, \qquad (3\text{-}10)$$

where T is a transformation matrix that maps the line l onto the x-axis, and R is the matrix that represents a reflection of the plane with respect to the x-axis (Figure 3-8).

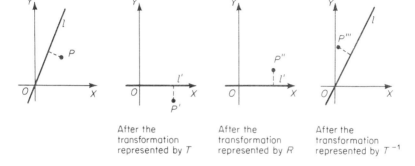

After the transformation represented by T

After the transformation represented by R

After the transformation represented by T^{-1}

Figure 3-8

In (3-8) and (3-9), transformation (3-10) has been applied in expressing the matrices that represent reflections of the plane with regard to the lines $y = x$ and $y = -x$, respectively. In (3-8), T is a matrix representing a rotation of the plane about the origin through an angle $\theta = -45°$, and in (3-9), T is a matrix representing a rotation of the plane about the origin through an angle $\theta = 45°$.

If T is a transformation matrix that maps the line l onto the y-axis, then in (3-10) R is the matrix that represents a reflection of the plane with respect to the y-axis. Another application of (3-10) is considered in the next example.

Example 1 Determine the reflection matrix F that maps each point (x, y) on the plane onto its mirror image with respect to the line $y = \sqrt{3}\,x$. Determine the image of the point $P: (\sqrt{3}, 1)$ under the reflection of the plane represented by F (Figure 3-9).

A rotation of the plane about the origin through an angle $\theta = 30°$ maps the line $y = \sqrt{3}\,x$ onto the y-axis. Such a rotation of the plane may be represented by the matrix

$$T = \begin{pmatrix} \frac{\sqrt{3}}{2} & -\frac{1}{2} \\ \frac{1}{2} & \frac{\sqrt{3}}{2} \end{pmatrix}, \quad \text{whose inverse} \quad T^{-1} = \begin{pmatrix} \frac{\sqrt{3}}{2} & \frac{1}{2} \\ -\frac{1}{2} & \frac{\sqrt{3}}{2} \end{pmatrix}.$$

The matrix R representing a reflection of the plane with respect to the y-axis is

$$\begin{pmatrix} -1 & 0 \\ 0 & 1 \end{pmatrix}.$$

Hence, by (3-10), the reflection matrix F that maps each point (x, y) on the plane onto its mirror image with respect to the line $y = \sqrt{3}\,x$ is given as

$$F = T^{-1}RT = \begin{pmatrix} \frac{\sqrt{3}}{2} & \frac{1}{2} \\ -\frac{1}{2} & \frac{\sqrt{3}}{2} \end{pmatrix} \begin{pmatrix} -1 & 0 \\ 0 & 1 \end{pmatrix} \begin{pmatrix} \frac{\sqrt{3}}{2} & -\frac{1}{2} \\ \frac{1}{2} & \frac{\sqrt{3}}{2} \end{pmatrix};$$

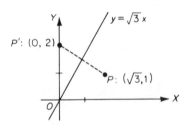

Figure 3-9

that is,

$$F = \begin{pmatrix} -\frac{1}{2} & \frac{\sqrt{3}}{2} \\ \frac{\sqrt{3}}{2} & \frac{1}{2} \end{pmatrix}.$$

The image of the point P: $(\sqrt{3}, 1)$ under the reflection of the plane represented by F is P': $(0, 2)$ since

$$\begin{pmatrix} -\frac{1}{2} & \frac{\sqrt{3}}{2} \\ \frac{\sqrt{3}}{2} & \frac{1}{2} \end{pmatrix} \begin{pmatrix} \sqrt{3} \\ 1 \end{pmatrix} = \begin{pmatrix} 0 \\ 2 \end{pmatrix}.$$

Another important class of transformations of the plane may be represented by the scalar matrices of order two

$$\begin{pmatrix} k & 0 \\ 0 & k \end{pmatrix},$$

where $k > 0$. The effect of such a transformation matrix is to map each point P on the plane onto a point P' on the plane such that $\overline{OP'} = k\overline{OP}$. If $k > 1$, the matrix represents a "uniform stretching" of the plane, and if $0 < k < 1$, the matrix represents a "uniform compression" of the plane. A transformation of the plane represented by a scalar matrix of order two having diagonal elements greater than zero is called a **dilation of the plane**.

Transformations represented by diagonal matrices of order two

$$\begin{pmatrix} d_1 & 0 \\ 0 & d_2 \end{pmatrix},$$

where $d_1 > 0$ and $d_2 > 0$, are related closely to the dilations of the plane. Many plane geometric figures are distorted under such transformations. A transformation of the plane represented by a diagonal matrix of order two with diagonal elements greater than zero is called a **magnification of the plane**. The set of dilations of the plane is a subset of the set of magnifications of the plane. Note that a magnification of the plane is a one-to-one mapping of the set of points on the plane onto itself.

Example 2 Determine the effect of the dilation of the plane represented by

$$\begin{pmatrix} 4 & 0 \\ 0 & 4 \end{pmatrix}$$

upon the line $y = 3x + 2$.

Under the dilation of the plane, each point (x, y) on the plane is mapped onto (x', y') where

$$\begin{pmatrix} x' \\ y' \end{pmatrix} = \begin{pmatrix} 4 & 0 \\ 0 & 4 \end{pmatrix} \begin{pmatrix} x \\ y \end{pmatrix}.$$

Hence,

$$\begin{pmatrix} x \\ y \end{pmatrix} = \begin{pmatrix} \frac{1}{4} & 0 \\ 0 & \frac{1}{4} \end{pmatrix} \begin{pmatrix} x' \\ y' \end{pmatrix};$$

that is, $x = \frac{1}{4}x'$ and $y = \frac{1}{4}y'$. Therefore, under the given dilation of the plane, the line $y = 3x + 2$ is mapped onto the line $y' = 3x' + 8$; that is, $y = 3x + 8$. Note that the image of the given line is a line with the same slope.

Example 3 Determine the effect of the magnification of the plane represented by

$$\begin{pmatrix} 1 & 0 \\ 0 & 3 \end{pmatrix}$$

upon the ellipse $\dfrac{x^2}{9} + \dfrac{y^2}{1} = 1$ (Figure 3–10).

Under the magnification of the plane, each point (x, y) on the plane is mapped onto (x', y') where

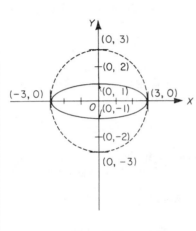

Figure 3-10

$$\begin{pmatrix} x' \\ y' \end{pmatrix} = \begin{pmatrix} 1 & 0 \\ 0 & 3 \end{pmatrix}\begin{pmatrix} x \\ y \end{pmatrix}.$$

Hence,

$$\begin{pmatrix} x \\ y \end{pmatrix} = \begin{pmatrix} 1 & 0 \\ 0 & \frac{1}{3} \end{pmatrix}\begin{pmatrix} x' \\ y' \end{pmatrix};$$

that is, $x = x'$ and $y = \frac{1}{3}y'$. Therefore, under the magnification of the plane represented by

$$\begin{pmatrix} 1 & 0 \\ 0 & 3 \end{pmatrix},$$

the ellipse $\dfrac{x^2}{9} + \dfrac{y^2}{1} = 1$ is mapped onto the circle $\dfrac{(x')^2}{9} + \dfrac{(y')^2}{9} = 1$; that is, $x^2 + y^2 = 9$.

Exercises

In Exercises 1 through 3 determine the effect upon the coordinate plane of the transformation of the plane represented by the matrix.

1. $\begin{pmatrix} 2 & 0 \\ 0 & 2 \end{pmatrix}.$ **2.** $\begin{pmatrix} -1 & 0 \\ 0 & -1 \end{pmatrix}.$ **3.** $\begin{pmatrix} 0 & 2 \\ 2 & 0 \end{pmatrix}.$

4. Prove that the products of the reflection matrices (3-6) represent rotations of the plane. Describe these rotations.

5. Prove that the distance between two points on a plane is invariant under a reflection of the plane with respect to the **(a)** x-axis; **(b)** y-axis.

6. Determine the reflection matrix F that maps each point (x, y) on the plane onto its mirror image with respect to the line $3x - 4y = 0$.

7. Prove that the multiplication of any dilation matrix and any rotation matrix of the form (3-3) is commutative.

8. Determine a matrix that represents the product of a reflection of the plane with respect to the line $y = x$ followed by a reflection of the plane with respect to the line $y = \sqrt{3}\,x$. Describe the resulting transformation matrix.

9. Find the dilation matrix that maps the points on a unit circle onto the points on a circle with radius (a) 4; (b) $\frac{1}{2}$; (c) r.

10. Determine the effect of the dilation of the plane represented by

$$\begin{pmatrix} 3 & 0 \\ 0 & 3 \end{pmatrix}$$

upon the triangle with vertices at $(0, 0)$, $(1, 2)$, and $(3, 1)$.

11. Determine the effect of the magnification of the plane represented by

$$\begin{pmatrix} \frac{1}{5} & 0 \\ 0 & \frac{1}{2} \end{pmatrix}$$

upon the line $2x + 5y = 10$.

12. Determine the effect of the magnification of the plane represented by

$$\begin{pmatrix} 2 & 0 \\ 0 & 1 \end{pmatrix}$$

upon the unit square with vertices at $(0, 0)$, $(1, 0)$, $(1, 1)$, and $(0, 1)$.

3-4 Other Transformations

The transformations of the plane discussed thus far have been examples of one-to-one mappings of the plane onto itself. The following examples present for consideration some other interesting transformations of the plane, two of which are not one-to-one mappings.

Example 1 The transformation of the plane represented by a matrix of the form

$$\begin{pmatrix} 1 & k \\ 0 & 1 \end{pmatrix}$$

is called a **shear parallel to the x-axis**. Determine the effect of the shear parallel to the x-axis represented by

$$\begin{pmatrix} 1 & 3 \\ 0 & 1 \end{pmatrix}$$

upon a rectangle with vertices at (0, 0), (2, 0), (2, 1), and (0, 1) (Figure 3-11).

Under the shear parallel to the x-axis represented by

$$\begin{pmatrix} 1 & 3 \\ 0 & 1 \end{pmatrix},$$

each point $(x, 0)$ is a fixed point; that is, each point on the x-axis is mapped onto itself. Hence, the line segment with end points $(0, 0)$ and $(2, 0)$ is mapped onto itself. Each point $(x, 1)$ is mapped onto a point with coordinates $(x + 3, 1)$; that is, each point on the line $y = 1$ is mapped onto another point of the line and positioned three units to the right. Hence, the line segment with end points $(0, 1)$ and $(2, 1)$ is mapped onto the line segment with end points $(3, 1)$ and $(5, 1)$. Each point $(0, y)$ is mapped onto a point with coordinates $(3y, y)$; that is, each point on the y-axis is mapped onto a point on the line $3y = x$. As a result the line segment with end points $(0, 0)$ and $(0, 1)$ is mapped onto the line segment with end points $(0, 0)$ and $(3, 1)$. Each point $(2, y)$ is mapped onto a point with coordinates $(2 + 3y, y)$; that is, each point on the line $x = 2$ is mapped

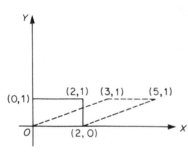

Figure 3-11

onto a point on the line $2 + 3y = x$. The result is that the line segment with end points $(2, 0)$ and $(2, 1)$ is mapped onto the line segment with end points $(2, 0)$ and $(5, 1)$. Hence, the rectangle is mapped onto a parallelogram with vertices at $(0, 0)$, $(2, 0)$, $(5, 1)$, and $(3, 1)$.

Example 2 Determine the effect of the transformation of the plane represented by

$$\begin{pmatrix} 1 & 0 \\ 0 & 0 \end{pmatrix}.$$

Under the transformation of the plane, each point (x, y) on the plane is mapped onto a point with coordinates $(x, 0)$. The matrix represents a vertical projection of the points on the plane onto the x-axis. The transformation of the plane represented by

$$\begin{pmatrix} 1 & 0 \\ 0 & 0 \end{pmatrix}$$

is an example of a mapping of the set of points on the plane into, but not onto, itself.

Example 3 Determine the effect of the transformation of the plane represented by

$$\begin{pmatrix} 1 & 0 \\ 1 & 0 \end{pmatrix}.$$

Under the transformation of the plane, each point (x, y) on the plane is mapped onto a point with coordinates (x, x). The matrix represents a projection of the plane onto the line $y = x$. The transformation of the plane represented by

$$\begin{pmatrix} 1 & 0 \\ 1 & 0 \end{pmatrix}$$

is an example of a mapping of the set of points on the plane into, but not onto, itself.

The transformations of the plane represented by the matrices of Examples 2 and 3 are called **projections of the plane**. Other projections of the plane exist.

Exercises

In Exercises 1 through 3 determine the effect upon the coordinate plane of the transformation of the plane represented by the matrix.

1. $\begin{pmatrix} 0 & 0 \\ 0 & 1 \end{pmatrix}.$ **2.** $\begin{pmatrix} 1 & 1 \\ 0 & 0 \end{pmatrix}.$ **3.** $\begin{pmatrix} 1 & -2 \\ 0 & 1 \end{pmatrix}.$

4. Determine the effect of the shear parallel to the x-axis represented by

$$\begin{pmatrix} 1 & 2 \\ 0 & 1 \end{pmatrix}$$

upon a rectangle with vertices at $(-1, -1)$, $(3, -1)$, $(3, 2)$, and $(-1, 2)$.

5. The transformation of the plane represented by a matrix of the form

$$\begin{pmatrix} 1 & 0 \\ k & 1 \end{pmatrix}$$

is called a **shear parallel to the y-axis**. (a) Determine the effect of the shear parallel to the y-axis represented by

$$\begin{pmatrix} 1 & 0 \\ 2 & 1 \end{pmatrix}$$

upon the unit circle $x^2 + y^2 = 1$. (**b**) Determine the effect of the shear parallel to the y-axis represented by

$$\begin{pmatrix} 1 & 0 \\ k & 1 \end{pmatrix}$$

upon a rectangle with vertices at $(0, 0)$, $(2, 0)$, $(2, 1)$, and $(0, 1)$.

6. Find a matrix that represents the product of a shear parallel to the x-axis followed by a dilation of the plane that maps the line $y = 3x - 2$ onto the line $x = 6$.

7. Determine a transformation matrix that maps the points $(1, 0)$ and $(0, 1)$ onto $(2, 1)$ and $(3, 2)$, respectively.

8. Prove that the multiplication of a matrix representing a shear parallel to the x-axis and a rotation matrix of the form (3-3) generally is not commutative.

3-5 Linear Homogeneous Transformations

The transformations of the plane that have been discussed thus far are examples of a general class of transformations called **linear homogeneous transformations of the plane**. These transformations always may be expressed in matrix form as

$$\begin{pmatrix} x' \\ y' \end{pmatrix} = \begin{pmatrix} a & b \\ c & d \end{pmatrix} \begin{pmatrix} x \\ y \end{pmatrix}; \tag{3-11}$$

that is, under a linear homogeneous transformation of the plane, each point (x, y) is mapped onto its image point (x', y') where

$$\begin{cases} x' = ax + by \\ y' = cx + dy. \end{cases} \tag{3-12}$$

If the matrix of the transformation

$$\begin{pmatrix} a & b \\ c & d \end{pmatrix}$$

is nonsingular, the transformation defined by (3-11) is called a **nonsingular linear homogeneous transformation of the plane.** For example, rotations of the plane about the origin, reflections of the plane with respect to a line through the origin, dilations of the plane, magnifications of the plane, and shears parallel to a coordinate axis are nonsingular linear homogeneous transformations of the plane. The projections of the plane considered in § 3-4 are linear homogeneous transformations of the plane that are not nonsingular.

Theorem 3-1 *Every nonsingular linear homogeneous transformation of the plane is a one-to-one mapping of the set of points on the plane onto itself.*

   Proof: Let T be a nonsingular matrix of order two. Consider,

$$\begin{pmatrix} x' \\ y' \end{pmatrix} = T\begin{pmatrix} x \\ y \end{pmatrix}.$$

Then, since T^{-1} exists,

$$T^{-1}\begin{pmatrix} x' \\ y' \end{pmatrix} = T^{-1}T\begin{pmatrix} x \\ y \end{pmatrix} = \begin{pmatrix} x \\ y \end{pmatrix};$$

that is, given any point P': (x', y'), it is possible to find the point P: (x, y) such that P' is the image of P under the transformation represented by T. Hence, each point on the plane is the image of some point on the plane under the transformation represented by T, and T represents a mapping of the set of points on the plane onto itself.

Let A: (x_1, y_1) and B: (x_2, y_2) be any two points on the plane that have the same image point under the transformation represented by T; that is, let

$$\begin{pmatrix} x' \\ y' \end{pmatrix} = T\begin{pmatrix} x_1 \\ y_1 \end{pmatrix} \quad \text{and} \quad \begin{pmatrix} x' \\ y' \end{pmatrix} = T\begin{pmatrix} x_2 \\ y_2 \end{pmatrix}.$$

Then

$$T\begin{pmatrix} x_1 \\ y_1 \end{pmatrix} = T\begin{pmatrix} x_2 \\ y_2 \end{pmatrix}$$

$$T^{-1}T\begin{pmatrix} x_1 \\ y_1 \end{pmatrix} = T^{-1}T\begin{pmatrix} x_2 \\ y_2 \end{pmatrix}$$

$$\begin{pmatrix} x_1 \\ y_1 \end{pmatrix} = \begin{pmatrix} x_2 \\ y_2 \end{pmatrix}.$$

Since A and B are identical points if they have the same image point, T represents a one-to-one mapping of the set of points on the plane onto itself.

One of the purposes for discussing reflections of the plane with respect to a line through the origin, magnifications of the plane, and shears parallel to a coordinate axis may be noted in the following interesting and important theorem.

Theorem 3-2 *Every nonsingular linear homogeneous transformation of the plane is a product of a reflection of the plane with respect to the line $y = x$, magnifications of the plane, and shears parallel to a coordinate axis.*

Proof: From the study in § 2-4 of elementary row transformations, it is known that every nonsingular matrix

$$\begin{pmatrix} a & b \\ c & d \end{pmatrix}$$

can be expressed as a product of the elementary row·transformation matrices of order two; that is,

$$\begin{pmatrix} a & b \\ c & d \end{pmatrix}$$

is a product of the matrices of the form

$$\begin{pmatrix} 0 & 1 \\ 1 & 0 \end{pmatrix}, \begin{pmatrix} k & 0 \\ 0 & 1 \end{pmatrix}, \begin{pmatrix} 1 & 0 \\ 0 & k \end{pmatrix}, \begin{pmatrix} 1 & k \\ 0 & 1 \end{pmatrix}, \text{ and } \begin{pmatrix} 1 & 0 \\ k & 1 \end{pmatrix}. \qquad (3\text{-}13)$$

The first matrix of (3-13) represents a reflection of the plane with respect to the line $y = x$; the second and third matrices represent magnifications of the plane; and the fourth and fifth matrices represent shears parallel to a coordinate axis.

Exercises

In Exercises 1 through 4 prove that under a nonsingular linear homogeneous transformation of the plane:

1. The image of the origin is the origin.

2. The image of a line is a line.

3. The images of parallel lines are parallel lines.

4. The image of a square is a parallelogram.

5. Determine the effect of the linear homogeneous transformation represented by

$$\begin{pmatrix} 6 & 3 \\ 2 & 1 \end{pmatrix}$$

upon the coordinate plane.

6. The set of points on the plane whose image points under a transformation of the plane are the origin is called the **null space** of the transformation. Find the null space of the transformation of the plane represented by

$$\begin{pmatrix} 4 & 2 \\ 2 & 1 \end{pmatrix}.$$

3-6 Orthogonal Matrices

A square real matrix A for which $AA^T = I$ is called an **orthogonal matrix**. Then A^T is the right inverse of an orthogonal matrix A. Since A is a square matrix its left and right inverses are equal. Hence, $A^T A = I$. Furthermore, $A^T = A^{-1}$; that is, a matrix is orthogonal if and only if its transpose and inverse are equal.

Let a_{ij} be the general element of an orthogonal matrix A of order n. Since

$$\sum_{k=1}^{n} a_{ik} a_{jk} = \delta_{ij},$$

the row vectors of A are mutually orthogonal unit vectors; that is, since the sum of the products of the elements of the ith row of A and the corresponding elements of the jth column of A^T taken in order is 0 if $i \neq j$, and 1 if $i = j$, and since the jth column of A^T is the jth row of A, any two rows of A represent orthogonal vectors, and any row vector is a unit vector. Similarly, it follows that the column vectors of an orthogonal matrix A are mutually orthogonal unit vectors since $A^T A = I$.

If A is an orthogonal matrix, $(A^{-1})(A^{-1})^T = A^T(A^T)^T = A^T A = I$; that is, the inverse (transpose) of an orthogonal matrix is an orthogonal matrix.

If two matrices A and B are orthogonal matrices, their product is an orthogonal matrix since $(AB)^T = B^T A^T = B^{-1} A^{-1} = (AB)^{-1}$.

If A is an orthogonal matrix, det $AA^T = $ det A det $A^T = $ det A det $A = 1$, and det $A = \pm 1$; that is, the determinant of an orthogonal matrix equals 1 or -1. If det $A = 1$, the orthogonal matrix A is called a **proper orthogonal matrix**; if det $A = -1$, the orthogonal matrix A is called an **improper orthogonal matrix**.

The transformation matrices that represent rotations of the plane about the origin are examples of proper orthogonal matrices since, in each case, the row (column) vectors are mutually orthogonal unit vectors and the value of the determinant of the matrix is 1. The transformation matrices that represent reflections of the plane with respect to a line through the origin are examples of improper orthogonal matrices.

Example 1 Verify that the matrix A is a proper orthogonal matrix where

$$A = \begin{pmatrix} \frac{12}{13} & \frac{5}{13} \\ -\frac{5}{13} & \frac{12}{13} \end{pmatrix}.$$

$$AA^T = \begin{pmatrix} \frac{12}{13} & \frac{5}{13} \\ -\frac{5}{13} & \frac{12}{13} \end{pmatrix} \begin{pmatrix} \frac{12}{13} & -\frac{5}{13} \\ \frac{5}{13} & \frac{12}{13} \end{pmatrix} = \begin{pmatrix} 1 & 0 \\ 0 & 1 \end{pmatrix}.$$

Since $AA^T = I$, A is an orthogonal matrix. Furthermore, det $A = 1$. Hence, A is a proper orthogonal matrix.

Since the matrices that represent rotations of the plane about the origin and reflections of the plane with respect to a line through the origin are orthogonal matrices, it is possible to present a concise proof which shows that the scalar product of two plane vectors is a scalar invariant under these transformations. That is, if $\vec{a}' = a_1' \vec{i} + a_2' \vec{j}$ and $\vec{b}' = b_1' \vec{i} + b_2' \vec{j}$ are the image vectors of $\vec{a} = a_1 \vec{i} + a_2 \vec{j}$ and $\vec{b} = b_1 \vec{i} + b_2 \vec{j}$, respectively, under a rotation of the plane about the origin or under a reflection of the plane with respect to a line through the origin, then $a_1 b_1 + a_2 b_2 = a_1' b_1' + a_2' b_2'$. Let R be either a rotation matrix or a reflection matrix. Then

$$(a_1'\quad a_2')^T = R(a_1\quad a_2)^T$$

and
$$(a_1\quad a_2)R^T = (a_1'\quad a_2').$$

Similarly,

$$R(b_1\quad b_2)^T = (b_1'\quad b_2')^T.$$

Then, multiplying equals by equals,

$$(a_1\quad a_2)R^T R(b_1\quad b_2)^T = (a_1'\quad a_2')(b_1'\quad b_2')^T.$$

Since R is an orthogonal matrix, $R^T R = I$. Hence,

$$(a_1\quad a_2)(b_1\quad b_2)^T = (a_1'\quad a_2')(b_1'\quad b_2')^T$$

and
$$a_1 b_1 + a_2 b_2 = a_1' b_1' + a_2' b_2'.$$

Since magnitudes of line segments, distances, and the measure of angles may be expressed in terms of the scalar product of two vectors, it immediately follows that these properties are invariants under rotations of the plane about the origin and under reflections of the plane with respect to a line through the origin.

Example 2 Verify that the scalar product of the vectors $\vec{a} = 2\vec{i} + \vec{j}$ and $\vec{b} = 3\vec{i} - \vec{j}$ is a scalar invariant under a rotation of the plane about the origin represented by

$$\begin{pmatrix} \frac{3}{5} & \frac{4}{5} \\ -\frac{4}{5} & \frac{3}{5} \end{pmatrix}.$$

Under the given rotation of the plane, $\vec{a}' = a_1'\vec{i} + a_2'\vec{j}$ where

$$\begin{pmatrix} a_1' \\ a_2' \end{pmatrix} = \begin{pmatrix} \frac{3}{5} & \frac{4}{5} \\ -\frac{4}{5} & \frac{3}{5} \end{pmatrix}\begin{pmatrix} 2 \\ 1 \end{pmatrix} = \begin{pmatrix} 2 \\ -1 \end{pmatrix}.$$

Similarly, $\vec{b}' = b_1'\vec{i} + b_2'\vec{j}$ where

$$\begin{pmatrix} b_1' \\ b_2' \end{pmatrix} = \begin{pmatrix} \frac{3}{5} & \frac{4}{5} \\ -\frac{4}{5} & \frac{3}{5} \end{pmatrix}\begin{pmatrix} 3 \\ -1 \end{pmatrix} = \begin{pmatrix} 1 \\ -3 \end{pmatrix}.$$

Then $\vec{a}\cdot\vec{b} = (2)(3) + (1)(-1) = 5$, and $\vec{a}'\cdot\vec{b}' = (2)(1) + (-1)(-3) = 5$. Hence, the scalar product $\vec{a}\cdot\vec{b}$ is a scalar invariant under the rotation of the plane about the origin represented by

$$\begin{pmatrix} \frac{3}{5} & \frac{4}{5} \\ -\frac{4}{5} & \frac{3}{5} \end{pmatrix}.$$

Exercises

In Exercises 1 through 4 determine whether the matrix is a proper orthogonal matrix or an improper orthogonal matrix.

1. $\begin{pmatrix} \frac{\sqrt{2}}{2} & \frac{\sqrt{2}}{2} \\ -\frac{\sqrt{2}}{2} & \frac{\sqrt{2}}{2} \end{pmatrix}.$
2. $\begin{pmatrix} \frac{\sqrt{3}}{2} & \frac{1}{2} \\ \frac{1}{2} & -\frac{\sqrt{3}}{2} \end{pmatrix}.$

3. $\begin{pmatrix} 2 & 0 \\ 1 & \frac{1}{2} \end{pmatrix}.$
4. $\begin{pmatrix} 3 & 5 \\ 1 & 2 \end{pmatrix}.$

5. Verify that the inverse of an orthogonal matrix A is an orthogonal matrix where

$$A = \begin{pmatrix} \frac{2}{3} & -\frac{2}{3} & \frac{1}{3} \\ \frac{1}{3} & \frac{2}{3} & \frac{2}{3} \\ \frac{2}{3} & \frac{1}{3} & -\frac{2}{3} \end{pmatrix}.$$

6. Verify that the product of two orthogonal matrices A and B is an orthogonal matrix where

$$A = \begin{pmatrix} -\frac{\sqrt{2}}{2} & -\frac{\sqrt{2}}{2} \\ \frac{\sqrt{2}}{2} & -\frac{\sqrt{2}}{2} \end{pmatrix} \quad \text{and} \quad B = \begin{pmatrix} \frac{3}{5} & \frac{4}{5} \\ \frac{4}{5} & -\frac{3}{5} \end{pmatrix}.$$

7. Verify that if A is an improper orthogonal matrix then $\det(A + I) = 0$ where

$$A = \begin{pmatrix} \frac{12}{13} & \frac{5}{13} \\ \frac{5}{13} & -\frac{12}{13} \end{pmatrix}.$$

8. Prove that the product of two improper orthogonal matrices of the same order is a proper orthogonal matrix.

9. Determine all the orthogonal matrices of order two whose elements are 0's and 1's.

10. Prove that if $AB = BA$ and C is an orthogonal matrix, then the multiplication of $C^T A C$ and $C^T B C$ is commutative.

3-7 Translations

The set of linear homogeneous transformations of the plane discussed in § 3-5 is a subset of the set of **general linear transformations of the plane**. Every general linear transformation of the plane may be described by the equations

$$\begin{cases} x' = ax + by + e \\ y' = cx + dy + f, \end{cases} \tag{3-14}$$

where (x', y') is the image of the point (x, y). It is not possible to express a general linear transformation of the plane described by (3-14) in matrix form where the matrix of the transformation is a square matrix of order two. To express (3-14) in matrix form it will be necessary to consider the use of homogeneous rectangular cartesian coordinates.

On a coordinate Euclidean plane, the **homogeneous coordinates** of a point with **nonhomogeneous coordinates** (x, y) are any three ordered scalars (x_1, x_2, x_3) where $x_3 \neq 0$ and for which $x = x_1/x_3$, and $y = x_2/x_3$. For example, the point on the plane with nonhomogeneous coordinates $(1, -3)$ may be represented by an infinite set of homogeneous coordinates of the form $(k, -3k, k)$ where k is any nonzero real number. One set of homogeneous coordinates for the point (x, y) is always of the form $(x, y, 1)$; all other sets of homogeneous coordinates are of the form (kx, ky, k) where $k \neq 0$. Generally, the context of a discussion will indicate whether homogeneous coordinates or nonhomogeneous coordinates are being considered.

Now, a general linear transformation of the plane described by (3-14) may be expressed in matrix form as

$$\begin{pmatrix} x' \\ y' \\ 1 \end{pmatrix} = \begin{pmatrix} a & b & e \\ c & d & f \\ 0 & 0 & 1 \end{pmatrix} \begin{pmatrix} x \\ y \\ 1 \end{pmatrix},$$
(3-15)

where $(x', y', 1)$ are the homogeneous coordinates of the image of the point with homogeneous coordinates $(x, y, 1)$.

Note that every linear homogeneous transformation of the plane may be expressed in the matrix form

$$\begin{pmatrix} x' \\ y' \\ 1 \end{pmatrix} = \begin{pmatrix} a & b & 0 \\ c & d & 0 \\ 0 & 0 & 1 \end{pmatrix} \begin{pmatrix} x \\ y \\ 1 \end{pmatrix}.$$
(3-16)

For example, a rotation of the plane about the origin through an angle θ may be expressed as

$$\begin{pmatrix} x' \\ y' \\ 1 \end{pmatrix} = \begin{pmatrix} \cos\theta & -\sin\theta & 0 \\ \sin\theta & \cos\theta & 0 \\ 0 & 0 & 1 \end{pmatrix} \begin{pmatrix} x \\ y \\ 1 \end{pmatrix}.$$

The matrix equation (3-16) represents the two equations of (3-12), which define a linear homogeneous transformation of the plane, and a trivial third equation, $1 = 1$, which does not place any conditions on the relationship between the coordinates of a point and those of its image point.

Consider a transformation of the

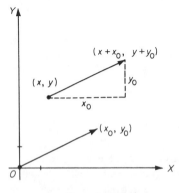

Figure 3-12

plane whereby each point on the plane is mapped onto a point a fixed directed distance x_0 away parallel to the x-axis and another fixed directed distance y_0 away parallel to the y-axis (Figure 3-12). Such a transformation of the plane is called a **translation of the plane** and is defined by the equations

$$\begin{cases} x' = x + x_0 \\ y' = y + y_0, \end{cases} \qquad (3\text{-}17)$$

where (x', y') is the image of the point (x, y). A translation of the plane may be expressed in matrix form as

$$\begin{pmatrix} x' \\ y' \\ 1 \end{pmatrix} = \begin{pmatrix} 1 & 0 & x_0 \\ 0 & 1 & y_0 \\ 0 & 0 & 1 \end{pmatrix} \begin{pmatrix} x \\ y \\ 1 \end{pmatrix}, \qquad (3\text{-}18)$$

where $(x', y', 1)$ are the homogeneous coordinates of the image of the point with homogeneous coordinates $(x, y, 1)$. The matrix

$$\begin{pmatrix} 1 & 0 & x_0 \\ 0 & 1 & y_0 \\ 0 & 0 & 1 \end{pmatrix} \qquad (3\text{-}19)$$

of the translation transformation defined by (3-18) is called a **translation matrix**. Every matrix representing a translation of the plane is of the form (3-19). Note that the determinant of every translation matrix is equal to one.

A translation of the plane is a one-to-one mapping of the set of points on the plane onto itself. Furthermore, there are no fixed points under the translation transformation (3-18) when x_0 and y_0 are not both zero; that is, each point is mapped onto another point. When $x_0 = y_0 = 0$, every point on the plane is a fixed point.

Example 1 Determine the coordinates of the image point of $P: (3, 2)$ under a translation of the plane that maps the origin onto $(4, -1)$.

The given translation of the plane may be described by the matrix equation

$$\begin{pmatrix} x' \\ y' \\ 1 \end{pmatrix} = \begin{pmatrix} 1 & 0 & 4 \\ 0 & 1 & -1 \\ 0 & 0 & 1 \end{pmatrix} \begin{pmatrix} x \\ y \\ 1 \end{pmatrix}$$

since

$$\begin{pmatrix} 4 \\ -1 \\ 1 \end{pmatrix} = \begin{pmatrix} 1 & 0 & 4 \\ 0 & 1 & -1 \\ 0 & 0 & 1 \end{pmatrix} \begin{pmatrix} 0 \\ 0 \\ 1 \end{pmatrix}.$$

Hence, the image of P: $(3, 2)$ is the point P': (x', y') where

$$\begin{pmatrix} x' \\ y' \\ 1 \end{pmatrix} = \begin{pmatrix} 1 & 0 & 4 \\ 0 & 1 & -1 \\ 0 & 0 & 1 \end{pmatrix} \begin{pmatrix} 3 \\ 2 \\ 1 \end{pmatrix} = \begin{pmatrix} 7 \\ 1 \\ 1 \end{pmatrix};$$

that is, P': $(7, 1)$ is the image point of P: $(3, 2)$ under the translation of the plane that maps the origin onto $(4, -1)$.

Example 2 Determine the equation satisfied by the set of image points of the locus of $x^2 + y^2 + 4x + 6y + 9 = 0$ under a translation of the plane represented by

$$\begin{pmatrix} 1 & 0 & 2 \\ 0 & 1 & 3 \\ 0 & 0 & 1 \end{pmatrix}.$$

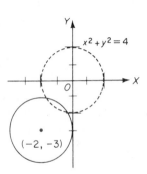

Figure 3-13

Under the given translation of the plane, each point (x, y) is mapped onto the point $(x + 2, y + 3)$. Hence, if x is replaced by $x - 2$ and y by $y - 3$ in $x^2 + y^2 + 4x + 6y + 9 = 0$, the equation satisfied by the set of image points becomes

$$(x - 2)^2 + (y - 3)^2 + 4(x - 2)$$
$$+ 6(y - 3) + 9 = 0,$$
$$x^2 - 4x + 4 + y^2 - 6y + 9$$
$$+ 4x - 8 + 6y - 18$$
$$+ 9 = 0,$$
$$x^2 + y^2 = 4;$$

that is, a circle with center at $(-2, -3)$ and a radius of 2 units is mapped onto a circle with center at the origin and a radius of 2 units (Figure 3-13).

Example 3 Show that the image of a point P: (x, y) under a rotation of the plane about the origin followed by a translation of the plane generally is not the image of the point under the same transformations considered in reverse order.

The matrices that represent a rotation of the plane about the origin and a translation of the plane are of the forms

$$\begin{pmatrix} \cos \theta & -\sin \theta & 0 \\ \sin \theta & \cos \theta & 0 \\ 0 & 0 & 1 \end{pmatrix} \text{ and } \begin{pmatrix} 1 & 0 & x_0 \\ 0 & 1 & y_0 \\ 0 & 0 & 1 \end{pmatrix},$$

respectively, if they are considered to be operating on P with coordinates expressed in homogeneous form. Then the coordinates of the image of P under a rotation followed by a translation are expressed by

$$\begin{pmatrix} x' \\ y' \\ 1 \end{pmatrix} = \begin{pmatrix} 1 & 0 & x_0 \\ 0 & 1 & y_0 \\ 0 & 0 & 1 \end{pmatrix} \begin{pmatrix} \cos\theta & -\sin\theta & 0 \\ \sin\theta & \cos\theta & 0 \\ 0 & 0 & 1 \end{pmatrix} \begin{pmatrix} x \\ y \\ 1 \end{pmatrix}$$
$$= \begin{pmatrix} x\cos\theta - y\sin\theta + x_0 \\ x\sin\theta + y\cos\theta + y_0 \\ 1 \end{pmatrix};$$

that is, (x, y) is mapped onto

$$(x\cos\theta - y\sin\theta + x_0,\ x\sin\theta + y\cos\theta + y_0).$$

The coordinates of the image of P under the same transformations considered in reverse order are expressed by

$$\begin{pmatrix} x' \\ y' \\ 1 \end{pmatrix} = \begin{pmatrix} \cos\theta & -\sin\theta & 0 \\ \sin\theta & \cos\theta & 0 \\ 0 & 0 & 1 \end{pmatrix} \begin{pmatrix} 1 & 0 & x_0 \\ 0 & 1 & y_0 \\ 0 & 0 & 1 \end{pmatrix} \begin{pmatrix} x \\ y \\ 1 \end{pmatrix}$$
$$= \begin{pmatrix} x\cos\theta - y\sin\theta + x_0\cos\theta - y_0\sin\theta \\ x\sin\theta + y\cos\theta + x_0\sin\theta + y_0\cos\theta \\ 1 \end{pmatrix};$$

that is, (x, y) is mapped onto

$$(x\cos\theta - y\sin\theta + x_0\cos\theta - y_0\sin\theta,$$
$$x\sin\theta + y\cos\theta + x_0\sin\theta + y_0\cos\theta).$$

In general, the two image points

$$(x\cos\theta - y\sin\theta + x_0,\ x\sin\theta + y\cos\theta + y_0)$$

and

$$(x\cos\theta - y\sin\theta + x_0\cos\theta - y_0\sin\theta,$$
$$x\sin\theta + y\cos\theta + x_0\sin\theta + y_0\cos\theta)$$

will not be the same.

Example 4 Use the results of Example 3 to determine the image of the point $P: (-5, 2)$ under a rotation of the plane about the origin through an angle of $90°$ followed by a translation of the plane that maps the origin onto $(2, 3)$.

Under the transformations taken in the given order, the image of each point (x, y) is

$$(x\cos\theta - y\sin\theta + x_0,\ x\sin\theta + y\cos\theta + y_0).$$

Since $\theta = 90°$, $x_0 = 2$, $y_0 = 3$, $x = -5$, and $y = 2$, the image of $P: (-5, 2)$ is $(-5 \cdot 0 - 2 \cdot 1 + 2, -5 \cdot 1 + 2 \cdot 0 + 3)$; that is, $(0, -2)$.

Exercises

1. Find the form of the homogeneous coordinates of the points having nonhomogeneous coordinates:
   (a) $(1, -2)$;   (b) $(3, 0)$;
   (c) $(0, 0)$;   (d) $(3, 4)$.

2. Find the nonhomogeneous coordinates of the points with homogeneous coordinates:
   (a) $(1, 3, 1)$;   (b) $(-2, 0, -1)$;
   (c) $(0, 4, 2)$;   (d) $(3, -6, 3)$.

3. Represent the following translations of the plane using matrices:
   (a) $x' = x - 2, y' = y + 4$;   (b) $x' = x, y' = y + 1$.

4. Determine the inverse of the translation matrix (3-19).

   In Exercises 5 through 7 determine the image point of the point P under a translation of the plane that maps the origin onto $(5, 2)$.

5. $(3, -4)$.   6. $(-5, -2)$.   7. $(5, 2)$.

8. Determine the equation satisfied by the set of image points of the locus of $xy + 4x - 3y - 13 = 0$ under a translation of the plane that maps the origin onto $(-3, 4)$.

9. Determine the equation satisfied by the set of image points of the locus of $3x^2 + 2y^2 - 6x + 8y - 7 = 0$ under a translation of the plane that maps the origin onto $(-1, 2)$.

10. Prove that the multiplication of translation matrices of the form (3-19) is (a) closed; (b) commutative.

11. Prove that the distance between two points on a plane is invariant under a translation of the plane.

12. Find the image of the point $P: (3 + \sqrt{2}, 4 - \sqrt{2})$ under a translation of the plane that maps the origin onto $(-3, -4)$ followed by a rotation of the plane about the new origin through an angle of $45°$.

13. Find the image of the point $P: \left(\dfrac{3 - 5\sqrt{3}}{2}, \dfrac{5 + \sqrt{3}}{2}\right)$ under a rotation of the plane about the origin through an angle of $30°$ followed by a translation of the plane that maps the origin onto $(2, -1)$.

3-8 Rigid Motion Transformations

In § 3-2 and § 3-3 rotations of the plane about the origin and reflections of the plane with respect to a line through the origin are considered. By means of the translation transformation and a generalization of the concept expressed by (3-10), it is now possible to consider rotations of the plane about any point on the plane and reflections of the plane with respect to any line on the plane. The following three examples illustrate the procedure for determining the matrices that represent such transformations.

Example 1 Determine the matrix $\mathcal{R}$ that represents a rotation of the plane about the point $(1, 1, 1)$ through an angle of $45°$. Determine the image of the origin under the rotation of the plane (Figure 3-14).

The translation of the plane that maps the point $(1, 1, 1)$ onto the origin is represented by the matrix

$$T = \begin{pmatrix} 1 & 0 & -1 \\ 0 & 1 & -1 \\ 0 & 0 & 1 \end{pmatrix}, \quad \text{whose inverse} \quad T^{-1} = \begin{pmatrix} 1 & 0 & 1 \\ 0 & 1 & 1 \\ 0 & 0 & 1 \end{pmatrix}.$$

The matrix

$$R = \begin{pmatrix} \frac{\sqrt{2}}{2} & -\frac{\sqrt{2}}{2} & 0 \\ \frac{\sqrt{2}}{2} & \frac{\sqrt{2}}{2} & 0 \\ 0 & 0 & 1 \end{pmatrix}$$

represents a rotation of the plane about the origin through an angle $\theta = 45°$. Hence, by a generalization of the concept expressed by (3-10), the matrix $\mathcal{R}$ that represents a rotation of the plane about the point $(1, 1, 1)$ through an angle of $45°$ is given as

$$\mathcal{R} = \begin{pmatrix} 1 & 0 & 1 \\ 0 & 1 & 1 \\ 0 & 0 & 1 \end{pmatrix} \begin{pmatrix} \frac{\sqrt{2}}{2} & -\frac{\sqrt{2}}{2} & 0 \\ \frac{\sqrt{2}}{2} & \frac{\sqrt{2}}{2} & 0 \\ 0 & 0 & 1 \end{pmatrix} \begin{pmatrix} 1 & 0 & -1 \\ 0 & 1 & -1 \\ 0 & 0 & 1 \end{pmatrix};$$

that is,

$$\mathcal{R} = \begin{pmatrix} \frac{\sqrt{2}}{2} & -\frac{\sqrt{2}}{2} & 1 \\ \frac{\sqrt{2}}{2} & \frac{\sqrt{2}}{2} & 1 - \sqrt{2} \\ 0 & 0 & 1 \end{pmatrix}.$$

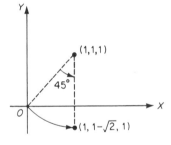

Figure 3-14

The image of the origin under the rotation of the plane represented by $\mathscr{R}$ is $(1, 1 - \sqrt{2}, 1)$ since

$$\begin{pmatrix} \frac{\sqrt{2}}{2} & -\frac{\sqrt{2}}{2} & 1 \\ \frac{\sqrt{2}}{2} & \frac{\sqrt{2}}{2} & 1 - \sqrt{2} \\ 0 & 0 & 1 \end{pmatrix} \begin{pmatrix} 0 \\ 0 \\ 1 \end{pmatrix} = \begin{pmatrix} 1 \\ 1 - \sqrt{2} \\ 1 \end{pmatrix}.$$

Example 2 Determine the reflection matrix F that maps each point $(x, y, 1)$ on the plane onto its mirror image with respect to the line $y = 2$.

The translation of the plane that maps the line $y = 2$ onto the x-axis is represented by the matrix

$$T = \begin{pmatrix} 1 & 0 & 0 \\ 0 & 1 & -2 \\ 0 & 0 & 1 \end{pmatrix}, \quad \text{whose inverse} \quad T^{-1} = \begin{pmatrix} 1 & 0 & 0 \\ 0 & 1 & 2 \\ 0 & 0 & 1 \end{pmatrix}.$$

The matrix

$$R = \begin{pmatrix} 1 & 0 & 0 \\ 0 & -1 & 0 \\ 0 & 0 & 1 \end{pmatrix}$$

represents a reflection of the plane with respect to the x-axis. Hence, by (3-10), the reflection matrix F that maps each point $(x, y, 1)$ on the plane onto its mirror image with respect to the line $y = 2$ is given as

$$F = T^{-1}RT = \begin{pmatrix} 1 & 0 & 0 \\ 0 & 1 & 2 \\ 0 & 0 & 1 \end{pmatrix} \begin{pmatrix} 1 & 0 & 0 \\ 0 & -1 & 0 \\ 0 & 0 & 1 \end{pmatrix} \begin{pmatrix} 1 & 0 & 0 \\ 0 & 1 & -2 \\ 0 & 0 & 1 \end{pmatrix};$$

that is,

$$F = \begin{pmatrix} 1 & 0 & 0 \\ 0 & -1 & 4 \\ 0 & 0 & 1 \end{pmatrix}.$$

Example 3 Determine the reflection matrix F that maps each point $(x, y, 1)$ on the plane onto its mirror image with respect to the line $x - y + 2 = 0$.

A translation of the plane represented by the matrix

$$\begin{pmatrix} 1 & 0 & 2 \\ 0 & 1 & 0 \\ 0 & 0 & 1 \end{pmatrix}$$

maps the line $x - y + 2 = 0$ onto the line $x - y = 0$; a rotation of the plane about the origin through an angle of $(-45°)$, represented by the matrix

$$\begin{pmatrix} \frac{\sqrt{2}}{2} & \frac{\sqrt{2}}{2} & 0 \\ -\frac{\sqrt{2}}{2} & \frac{\sqrt{2}}{2} & 0 \\ 0 & 0 & 1 \end{pmatrix},$$

maps the line $x - y = 0$ onto the x-axis. Hence, the product transformation that maps the line $x - y + 2 = 0$ onto the x-axis is represented by the matrix

$$T = \begin{pmatrix} \frac{\sqrt{2}}{2} & \frac{\sqrt{2}}{2} & 0 \\ -\frac{\sqrt{2}}{2} & \frac{\sqrt{2}}{2} & 0 \\ 0 & 0 & 1 \end{pmatrix} \begin{pmatrix} 1 & 0 & 2 \\ 0 & 1 & 0 \\ 0 & 0 & 1 \end{pmatrix} = \begin{pmatrix} \frac{\sqrt{2}}{2} & \frac{\sqrt{2}}{2} & \sqrt{2} \\ -\frac{\sqrt{2}}{2} & \frac{\sqrt{2}}{2} & -\sqrt{2} \\ 0 & 0 & 1 \end{pmatrix},$$

whose inverse

$$T^{-1} = \begin{pmatrix} 1 & 0 & -2 \\ 0 & 1 & 0 \\ 0 & 0 & 1 \end{pmatrix} \begin{pmatrix} \frac{\sqrt{2}}{2} & -\frac{\sqrt{2}}{2} & 0 \\ \frac{\sqrt{2}}{2} & \frac{\sqrt{2}}{2} & 0 \\ 0 & 0 & 1 \end{pmatrix} = \begin{pmatrix} \frac{\sqrt{2}}{2} & -\frac{\sqrt{2}}{2} & -2 \\ \frac{\sqrt{2}}{2} & \frac{\sqrt{2}}{2} & 0 \\ 0 & 0 & 1 \end{pmatrix}.$$

The matrix

$$R = \begin{pmatrix} 1 & 0 & 0 \\ 0 & -1 & 0 \\ 0 & 0 & 1 \end{pmatrix}$$

represents a reflection of the plane with respect to the x-axis. Hence, by (3-10), the reflection matrix F, which maps each point $(x, y, 1)$ on the plane onto its mirror image with respect to the line $x - y + 2 = 0$, is given as

$$F = T^{-1}RT = \begin{pmatrix} \frac{\sqrt{2}}{2} & -\frac{\sqrt{2}}{2} & -2 \\ \frac{\sqrt{2}}{2} & \frac{\sqrt{2}}{2} & 0 \\ 0 & 0 & 1 \end{pmatrix} \begin{pmatrix} 1 & 0 & 0 \\ 0 & -1 & 0 \\ 0 & 0 & 1 \end{pmatrix} \begin{pmatrix} \frac{\sqrt{2}}{2} & \frac{\sqrt{2}}{2} & \sqrt{2} \\ -\frac{\sqrt{2}}{2} & \frac{\sqrt{2}}{2} & -\sqrt{2} \\ 0 & 0 & 1 \end{pmatrix};$$

that is,

$$F = \begin{pmatrix} 0 & 1 & -2 \\ 1 & 0 & 2 \\ 0 & 0 & 1 \end{pmatrix}.$$

Note that under the transformation of the plane represented by F every point on the line $x - y + 2 = 0$ is a fixed point; that is,

$$\begin{pmatrix} x \\ x + 2 \\ 1 \end{pmatrix} = \begin{pmatrix} 0 & 1 & -2 \\ 1 & 0 & 2 \\ 0 & 0 & 1 \end{pmatrix} \begin{pmatrix} x \\ x + 2 \\ 1 \end{pmatrix}.$$

The set of rotations of the plane about a point, reflections of the plane with respect to a line, and translations of the plane is called the set of **rigid**

General linear transformations

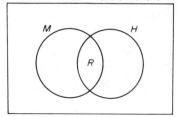

Figure 3-15

motion transformations because the distance between any two points on the plane is a scalar invariant under these transformations. (Euclidean geometry is sometimes characterized as a study of the properties of geometric figures which remain invariant under the rigid motion transformations.) The set of rigid motion transformations is a subset of the set of general linear transformations of the plane. As indicated in Figure 3-15, the set R of rotations of the plane about the origin and reflections of the plane with respect to a line through the origin is the intersection of the set M of rigid motion transformations and the set H of linear homogeneous transformations.

The set of ordered products of the rigid motion transformations of the plane may be represented by a matrix of the form

$$\begin{pmatrix} a_{11} & a_{12} & a_{13} \\ a_{21} & a_{22} & a_{23} \\ 0 & 0 & 1 \end{pmatrix}, \quad \text{where} \quad \begin{vmatrix} a_{11} & a_{12} \\ a_{21} & a_{22} \end{vmatrix} = \pm 1. \tag{3-20}$$

Furthermore,

$$\sum_{i=1}^{2} a_{ij}^2 = 1, \quad \text{for} \quad j = 1, 2; \tag{3-21}$$

and

$$\sum_{i=1}^{2} a_{i1} a_{i2} = 0. \tag{3-22}$$

Two theorems which state that a general linear transformation of the plane is a rigid motion transformation if and only if the matrix representing the transformation satisfies conditions (3-21) and (3-22) will be proved now.

Theorem 3-3 *Let*

$$T = \begin{pmatrix} a_{11} & a_{12} & a_{13} \\ a_{21} & a_{22} & a_{23} \\ 0 & 0 & 1 \end{pmatrix}$$

represent a general linear transformation of the plane under which distance is a scalar invariant. Then

$$a_{11}^2 + a_{21}^2 = 1,$$

$$a_{12}^2 + a_{22}^2 = 1,$$

and

$$a_{11}a_{12} + a_{21}a_{22} = 0.$$

Proof: Consider the points O: $(0, 0, 1)$ and P: $(1, 0, 1)$. Under the transformation represented by T, the image points of O and P are O': $(a_{13}, a_{23}, 1)$ and P': $(a_{11} + a_{13}, a_{21} + a_{23}, 1)$, respectively. Since $|\overrightarrow{O'P'}| = |\overrightarrow{OP}|$ under the transformation represented by T,

$$\sqrt{[(a_{11} + a_{13}) - a_{13}]^2 + [(a_{21} + a_{23}) - a_{23}]^2} = \sqrt{(1 - 0)^2 + (0 - 0)^2}$$

$$\sqrt{a_{11}^2 + a_{21}^2} = \sqrt{1^2}$$

$$a_{11}^2 + a_{21}^2 = 1.$$

In a similar manner, choosing O: $(0, 0, 1)$ and P: $(0, 1, 1)$, it can be shown that

$$a_{12}^2 + a_{22}^2 = 1.$$

Now, consider the points O: $(0, 0, 1)$ and P: $(1, 1, 1)$. Under the transformation represented by T, the image points of O and P are O': $(a_{13}, a_{23}, 1)$ and P': $(a_{11} + a_{12} + a_{13}, a_{21} + a_{22} + a_{23}, 1)$, respectively. Again, since $|\overrightarrow{O'P'}| = |\overrightarrow{OP}|$ under the transformation represented by T,

$$\sqrt{(a_{11} + a_{12})^2 + (a_{21} + a_{22})^2} = \sqrt{1^2 + 1^2}$$

$$(a_{11} + a_{12})^2 + (a_{21} + a_{22})^2 = 2$$

$$(a_{11}^2 + a_{21}^2) + (a_{12}^2 + a_{22}^2) + 2(a_{11}a_{12} + a_{21}a_{22}) = 2$$

$$1 + 1 + 2(a_{11}a_{12} + a_{21}a_{22}) = 2$$

$$a_{11}a_{12} + a_{21}a_{22} = 0.$$

Theorem 3-4 *Let*

$$T = \begin{pmatrix} a_{11} & a_{12} & a_{13} \\ a_{21} & a_{22} & a_{23} \\ 0 & 0 & 1 \end{pmatrix}$$

represent a general linear transformation of the plane such that $a_{11}^2 + a_{21}^2 = 1$, $a_{12}^2 + a_{22}^2 = 1$, and $a_{11}a_{12} + a_{21}a_{22} = 0$. Then distance is a scalar invariant under the transformation represented by T.

Proof: Consider any two points P_1: $(x_1, y_1, 1)$ and P_2: $(x_2, y_2, 1)$ on the plane. Under the transformation represented by T, the image points of P_1 and P_2 are P_1': $(a_{11}x_1 + a_{12}y_1 + a_{13}, a_{21}x_1 + a_{22}y_1 + a_{23}, 1)$ and P_2': $(a_{11}x_2 + a_{12}y_2 + a_{13}, a_{21}x_2 + a_{22}y_2 + a_{23}, 1)$, respectively. Now,

$$|\overrightarrow{P_1'P_2'}| = \sqrt{[a_{11}(x_2 - x_1) + a_{12}(y_2 - y_1)]^2 + [a_{21}(x_2 - x_1) + a_{22}(y_2 - y_1)]^2}$$

$$= [(a_{11}^2 + a_{21}^2)(x_2 - x_1)^2 + 2(a_{11}a_{12} + a_{21}a_{22})(x_2 - x_1)(y_2 - y_1)$$

$$+ (a_{12}^2 + a_{22}^2)(y_2 - y_1)^2]^{1/2}$$

$$= \sqrt{(x_2 - x_1)^2 + (y_2 - y_1)^2}$$

$$= |\overrightarrow{P_1P_2}|;$$

that is, the distance between the image points of P_1 and P_2 is equal to the distance between P_1 and P_2. Hence, distance is a scalar invariant under the transformation represented by T.

Exercises

1. Determine the matrix $\mathscr{R}$ that represents a rotation of the plane about the point $(1, \sqrt{3}, 1)$ through an angle of $60°$. Determine the image of the origin under the rotation of the plane.

2. Determine the matrix $\mathscr{R}$ that represents a rotation of the plane about the point $(1, 0, 1)$ through an angle of $90°$. Determine the image of the origin under the rotation of the plane.

3. Verify that the matrix of Exercise 1 is of the form (3-20) and satisfies the conditions (3-21) and (3-22).

4. Verify that the matrix of Exercise 2 is of the form (3-20) and satisfies the conditions (3-21) and (3-22).

5. Determine the reflection matrix F that maps each point $(x, y, 1)$ on the plane onto its mirror image with respect to the line **(a)** $x = 3$; **(b)** $x - \sqrt{3}\,y + 1 = 0$.

6. Verify that the product of a matrix representing a rotation of the plane about the origin and a matrix representing a translation of the plane, taken in either order, is of the form (3-20) and satisfies the conditions (3-21) and (3-22).

7. Verify that the product of a matrix representing a rotation of the plane about the origin and a matrix representing a reflection of the plane about a coordinate axis, taken in either order, is of the form (3-20) and satisfies the conditions (3-21) and (3-22).

chapter 4

Eigenvalues and Eigenvectors

4-1 Characteristic Functions

Associated with each square matrix $A = ((a_{ij}))$ of order n is a function

$$f(\lambda) = |A - \lambda I| = \begin{vmatrix} a_{11} - \lambda & a_{12} & \cdots & a_{1n} \\ a_{21} & a_{22} - \lambda & \cdots & a_{2n} \\ \cdots & \cdots & \cdots & \cdots \\ a_{n1} & a_{n2} & \cdots & a_{nn} - \lambda \end{vmatrix} \qquad (4\text{-}1)$$

called the **characteristic function** of A. The equation

$$f(\lambda) = |A - \lambda I| = 0 \qquad (4\text{-}2)$$

can be expressed in the polynomial form

$$c_0 \lambda^n + c_1 \lambda^{n-1} + \cdots + c_{n-1} \lambda + c_n = 0 \qquad (4\text{-}3)$$

and is called the **characteristic equation** of matrix A.

Example 1 Find the characteristic equation of matrix A where

$$A = \begin{pmatrix} 1 & 2 & 0 \\ 2 & 2 & 2 \\ 0 & 2 & 3 \end{pmatrix}.$$

The characteristic equation of A is

$$\begin{vmatrix} 1 - \lambda & 2 & 0 \\ 2 & 2 - \lambda & 2 \\ 0 & 2 & 3 - \lambda \end{vmatrix} = 0;$$

that is,

$$(1 - \lambda) \begin{vmatrix} 2 - \lambda & 2 \\ 2 & 3 - \lambda \end{vmatrix} - 2 \begin{vmatrix} 2 & 2 \\ 0 & 3 - \lambda \end{vmatrix} = 0$$

$$(1 - \lambda)(\lambda^2 - 5\lambda + 2) - 2(6 - 2\lambda) = 0$$

$$\lambda^3 - 6\lambda^2 + 3\lambda + 10 = 0.$$

In some instances the task of expressing the characteristic equation of a matrix in polynomial form may be simplified considerably by introducing the concept of the trace of a matrix. The sum of the diagonal elements of a matrix A is called the **trace** of A and is denoted by $tr(A)$. For example, the trace of matrix A in Example 1 is $1 + 2 + 3$; that is, 6. Let $t_1 = tr(A)$, $t_2 = tr(A^2), \ldots, t_n = tr(A^n)$. It can be shown that the coefficients of the characteristic equation are given by the equations:

$$c_0 = 1,$$

$$c_1 = -t_1,$$

$$c_2 = -\tfrac{1}{2}(c_1 t_1 + t_2),$$

$$c_3 = -\tfrac{1}{3}(c_2 t_1 + c_1 t_2 + t_3), \tag{4-4}$$

$$\cdots \quad \cdots$$

$$c_n = -\frac{1}{n}(c_{n-1} t_1 + c_{n-2} t_2 + \cdots + c_1 t_{n-1} + t_n).$$

Equations (4-4) make it possible to calculate the coefficients of the characteristic equation of a matrix A by summing the diagonal elements of the matrices of the form A^n. This numerical process is easily programmed on a large-scale digital computer, or for small values of n may be computed manually without difficulty.

Example 2 Find the characteristic equation of matrix A in Example 1 by using the equations of (4-4).

The characteristic equation of A is of the form $c_0 \lambda^3 + c_1 \lambda^2 + c_2 \lambda + c_3 = 0$. Now,

$$t_1 = tr(A) = 1 + 2 + 3 = 6,$$

and since

$$A^2 = \begin{pmatrix} 5 & 6 & 4 \\ 6 & 12 & 10 \\ 4 & 10 & 13 \end{pmatrix} \quad \text{and} \quad A^3 = \begin{pmatrix} 17 & 30 & 24 \\ 30 & 56 & 54 \\ 24 & 54 & 59 \end{pmatrix},$$

$$t_2 = tr(A^2) = 5 + 12 + 13 = 30,$$
$$t_3 = tr(A^3) = 17 + 56 + 59 = 132.$$

Then, using the equations of (4-4),

$$c_0 = 1,$$
$$c_1 = -6,$$
$$c_2 = -\tfrac{1}{2}[(-6)(6) + 30] = 3,$$
$$c_3 = -\tfrac{1}{3}[(3)(6) + (-6)(30) + 132] = 10.$$

Hence, $\lambda^3 - 6\lambda^2 + 3\lambda + 10 = 0$ is the characteristic equation of A.

The n roots $\lambda_1, \lambda_2, \ldots, \lambda_n$ of the characteristic equation (4-3) of a matrix A are called the **eigenvalues** of A.

Example 3 Determine the eigenvalues of the matrix A where

$$A = \begin{pmatrix} 3 & 1 \\ 2 & 2 \end{pmatrix}.$$

The characteristic equation of A is

$$\begin{vmatrix} 3 - \lambda & 1 \\ 2 & 2 - \lambda \end{vmatrix} = 0.$$

In expanded form this becomes $\lambda^2 - 5\lambda + 4 = 0$. The eigenvalues $\lambda_1 = 1$ and $\lambda_2 = 4$ are the roots of the characteristic equation.

Example 4 Prove that the trace of a matrix A of order n is equal to the sum of the n eigenvalues of A; that is,

$$tr(A) = \sum_{i=1}^{n} \lambda_i.$$

Let $A = ((a_{ij}))$. By definition,

$$tr(A) = \sum_{i=1}^{n} a_{ii}.$$

Consider the characteristic function of A expressed in factored form:

$$|A - \lambda I| = (-1)^n (\lambda - \lambda_1)(\lambda - \lambda_2) \cdots (\lambda - \lambda_n).$$

The coefficient of λ^{n-1} in $|A - \lambda I|$ is

$$(-1)^{n-1} \sum_{i=1}^{n} a_{ii};$$

the coefficient of λ^{n-1} in the factored form of the characteristic function is

$$(-1)^{n+1} \sum_{i=1}^{n} \lambda_i.$$

Therefore,

$$(-1)^{n-1} \sum_{i=1}^{n} a_{ii} = (-1)^{n+1} \sum_{i=1}^{n} \lambda_i$$

$$\sum_{i=1}^{n} a_{ii} = \sum_{i=1}^{n} \lambda_i;$$

that is,
$$tr(A) = \sum_{i=1}^{n} \lambda_i.$$

Many applications of matrix algebra in mathematics, physics, and engineering involve the concept of a set of nonzero vectors being mapped onto the zero vector by means of the matrix $A - \lambda_i I$, where λ_i is an eigenvalue of matrix A. Any nonzero column vector, denoted by X_i, such that

$$(A - \lambda_i I)X_i = 0 \tag{4-5}$$

is called an **eigenvector** of matrix A. It is guaranteed that at least one eigenvector exists for each λ_i since equation (4-5) represents a system of n linear homogeneous equations which has a nontrivial solution $X_i \neq 0$ if and only if $|A - \lambda_i I| = 0$; that is, if and only if λ_i is an eigenvalue of A. Furthermore, note that any nonzero scalar multiple of an eigenvector associated with an eigenvalue is also an eigenvector associated with that eigenvalue.

Example 5 Determine a set of eigenvectors of the matrix A in Example 3.

Associated with $\lambda_1 = 1$ are the eigenvectors $(x_1 \quad x_2)^T$ for which
$$(A - I)(x_1 \quad x_2)^T = 0;$$
that is,
$$\begin{pmatrix} 2 & 1 \\ 2 & 1 \end{pmatrix}\begin{pmatrix} x_1 \\ x_2 \end{pmatrix} = \begin{pmatrix} 0 \\ 0 \end{pmatrix}.$$

It follows that $-2x_1 = x_2$. If x_1 is chosen as some convenient arbitrary scalar, say 1, x_2 becomes -2. Hence, $(1 \quad -2)^T$ is an eigenvector associated with the eigenvalue 1.

Similarly, associated with $\lambda_2 = 4$ are the eigenvectors $(x_1 \quad x_2)^T$ for which
$$(A - 4I)(x_1 \quad x_2)^T = 0;$$
that is,
$$\begin{pmatrix} -1 & 1 \\ 2 & -2 \end{pmatrix}\begin{pmatrix} x_1 \\ x_2 \end{pmatrix} = \begin{pmatrix} 0 \\ 0 \end{pmatrix}.$$

Hence, $x_1 = x_2$, and $(1 \quad 1)^T$ is an eigenvector associated with the eigenvalue 4. Therefore, one set of eigenvectors of the matrix A is $\{(1 \quad -2)^T, (1 \quad 1)^T\}$.

It should be noted that $(k \quad -2k)^T$ and $(k \quad k)^T$, where k is any nonzero scalar, represent the general forms of the eigenvectors of A.

The eigenvalues of a matrix are also called the **proper values**, the **latent values**, and the **characteristic values** of the matrix. The eigenvectors of a matrix are also called the **proper vectors**, the **latent vectors**, and the **characteristic vectors** of the matrix.

Exercises

In Exercises 1 through 6 determine the characteristic equation, the eigenvalues, and a set of eigenvectors of the given matrix.

1. $\begin{pmatrix} 5 & 3 \\ 2 & 4 \end{pmatrix}.$ **2.** $\begin{pmatrix} 1 & 2 \\ 4 & 3 \end{pmatrix}.$ **3.** $\begin{pmatrix} 2 & 0 \\ 0 & 0 \end{pmatrix}.$

4. $\begin{pmatrix} 2 & 0 & 0 \\ 0 & 1 & 0 \\ 0 & 0 & 3 \end{pmatrix}.$ **5.** $\begin{pmatrix} 2 & -2 & 3 \\ 1 & 1 & 1 \\ 1 & 3 & -1 \end{pmatrix}.$ **6.** $\begin{pmatrix} 3 & 0 & 2 \\ 0 & 1 & 2 \\ 2 & 2 & 2 \end{pmatrix}.$

7. Use the equations of (4-4) to find the characteristic equation of the matrix in **(a)** Exercise 1; **(b)** Exercise 5.

8. Verify the results of Example 4 for the matrix in **(a)** Exercise 1; **(b)** Exercise 6.

9. Prove that if an eigenvalue of A is zero, then $\det A = 0$.

10. Determine the necessary and sufficient conditions for a symmetric matrix of order two to have distinct eigenvalues.

11. If λ_1, λ_2, and λ_3 are the eigenvalues of a matrix A of order three, find the eigenvalues of **(a)** kA; **(b)** $A - kI$.

12. Prove that the eigenvalues of A and A^T are identical.

13. Prove that the eigenvalues of a diagonal matrix are equal to the diagonal elements.

4-2 A Geometric Interpretation of Eigenvectors

Consider a magnification of the plane represented by the matrix A where

$$A = \begin{pmatrix} 3 & 0 \\ 0 & 2 \end{pmatrix}.$$

The eigenvalues of A are $\lambda_1 = 3$ and $\lambda_2 = 2$. Every eigenvector associated with λ_1 is of the form $(k \quad 0)^T$, where k is any nonzero scalar, since

$$\begin{pmatrix} 3-3 & 0 \\ 0 & 2-3 \end{pmatrix}\begin{pmatrix} k \\ 0 \end{pmatrix} = \begin{pmatrix} 0 \\ 0 \end{pmatrix}.$$

Furthermore, the set of vectors of the form $(k \quad 0)^T$ is such that

$$A(k \quad 0)^T = \lambda_1(k \quad 0)^T;$$

that is,

$$\begin{pmatrix} 3 & 0 \\ 0 & 2 \end{pmatrix}\begin{pmatrix} k \\ 0 \end{pmatrix} = 3\begin{pmatrix} k \\ 0 \end{pmatrix}.$$

Hence, the set of eigenvectors associated with $\lambda_1 = 3$ is mapped onto itself under the transformation represented by A, and the image of each eigenvector is a fixed scalar multiple of that eigenvector. The fixed scalar multiple is equal to the eigenvalue with which the set of eigenvectors is associated.

Similarly, every eigenvector associated with λ_2 is of the form $(0 \quad k)^T$, where k is any nonzero scalar. The set of vectors of the form $(0 \quad k)^T$ is such that

$$A(0 \quad k)^T = \lambda_2(0 \quad k)^T;$$

that is,

$$\begin{pmatrix} 3 & 0 \\ 0 & 2 \end{pmatrix}\begin{pmatrix} 0 \\ k \end{pmatrix} = 2\begin{pmatrix} 0 \\ k \end{pmatrix}.$$

Hence, the set of eigenvectors associated with $\lambda_2 = 2$ is mapped onto itself under the transformation represented by A, and the image of each eigenvector is a fixed scalar multiple of the eigenvalue. The fixed scalar multiple is λ_2; that is, 2.

Note that the sets of vectors of the forms $(k \quad 0)^T$ and $(0 \quad k)^T$ lie along the x-axis and y-axis, respectively (Figure 4-1). Under the magnification

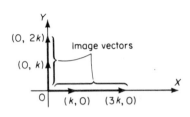

Figure 4-1

of the plane represented by the matrix

$$A = \begin{pmatrix} 3 & 0 \\ 0 & 2 \end{pmatrix},$$

the *one-dimensional vector spaces* containing the sets of vectors of the forms $(k \quad 0)^T$ and $(0 \quad k)^T$ are mapped *onto* themselves, respectively, and are called **invariant vector spaces**. The invariant vector spaces help characterize or describe a particular transformation of the plane.

Example 1 Determine the invariant vector spaces under a shear parallel to the x-axis represented by the matrix A where

$$A = \begin{pmatrix} 1 & 2 \\ 0 & 1 \end{pmatrix}.$$

The eigenvalues of A are $\lambda_1 = 1$ and $\lambda_2 = 1$. Associated with each eigenvalue is the set of eigenvectors of the form $(k \quad 0)^T$, where k is any nonzero scalar. Then

$$\begin{pmatrix} 1 & 2 \\ 0 & 1 \end{pmatrix}\begin{pmatrix} k \\ 0 \end{pmatrix} = 1\begin{pmatrix} k \\ 0 \end{pmatrix},$$

and the one-dimensional vector space containing the set of vectors of the form $(k \quad 0)^T$ is an invariant vector space. Furthermore, since $\lambda_1 = \lambda_2 = 1$, each vector in the vector space is its own image.

Example 2 Determine the invariant vector spaces under a projection of the plane represented by the matrix A where

$$A = \begin{pmatrix} 1 & 0 \\ 1 & 0 \end{pmatrix}.$$

The eigenvalues of A are $\lambda_1 = 1$ and $\lambda_2 = 0$. Associated with the eigenvalue $\lambda_1 = 1$ is the set of eigenvectors of the form $(k \quad k)^T$, where k is any nonzero scalar; associated with the eigenvalue $\lambda_2 = 0$ is the set of eigenvectors of the form $(0 \quad k)^T$, where k is any nonzero scalar. Then

$$\begin{pmatrix} 1 & 0 \\ 1 & 0 \end{pmatrix}\begin{pmatrix} k \\ k \end{pmatrix} = 1\begin{pmatrix} k \\ k \end{pmatrix}$$

and

$$\begin{pmatrix} 1 & 0 \\ 1 & 0 \end{pmatrix}\begin{pmatrix} 0 \\ k \end{pmatrix} = 0\begin{pmatrix} 0 \\ k \end{pmatrix} = \begin{pmatrix} 0 \\ 0 \end{pmatrix}.$$

Since the vectors of the form $(0 \quad k)^T$ are mapped onto the zero vector, the one-dimensional vector space containing these vectors is mapped *into* but not *onto* itself and the space is *not* considered an invariant vector space. However, the one-dimensional vector space containing the set of vectors of the form $(k \quad k)^T$ is an invariant vector space. Note that these vectors lie along the line $y = x$ and that the plane is mapped onto this line under the projection of the plane represented by matrix A.

Exercises

In Exercises 1 through 6 determine the invariant vector spaces under the transformation of the plane represented by the given matrix.

1. $\begin{pmatrix} 1 & 0 \\ 0 & -1 \end{pmatrix}.$

2. $\begin{pmatrix} 1 & 0 \\ 0 & 2 \end{pmatrix}.$

3. $\begin{pmatrix} 1 & 0 \\ 3 & 1 \end{pmatrix}.$

4. $\begin{pmatrix} 0 & 1 \\ 1 & 0 \end{pmatrix}.$

5. $\begin{pmatrix} 1 & 0 \\ 0 & 0 \end{pmatrix}.$

6. $\begin{pmatrix} 2 & 0 \\ 0 & 2 \end{pmatrix}.$

7. Prove that, in general, no invariant vector spaces exist under a rotation of the plane about the origin.

4-3 Some Theorems

In this section several theorems concerning the eigenvalues and eigenvectors of matrices in general and of symmetric matrices in particular will be proved. These theorems are important for an understanding of the remaining sections of this text.

Notice that in Example 5 of § 4-1 the eigenvectors associated with the distinct eigenvalues of matrix A are linearly independent; that is,

$$k_1 \begin{pmatrix} 1 \\ -2 \end{pmatrix} + k_2 \begin{pmatrix} 1 \\ 1 \end{pmatrix} = \begin{pmatrix} 0 \\ 0 \end{pmatrix}$$

implies $k_1 = k_2 = 0$. This is not a coincidence. The following theorem states a sufficient condition for eigenvectors associated with the eigenvalues of a matrix to be linearly independent.

Theorem 4-1 *If the eigenvalues of a matrix are distinct, then the associated eigenvectors are linearly independent.*

Proof: Let A be a square matrix of order n with distinct eigenvalues $\lambda_1, \lambda_2, \ldots, \lambda_n$ and associated eigenvectors $X_1, X_2, \ldots, X_n$, respectively. Assume that the set of eigenvectors are linearly dependent. Then there exists scalars $k_1, k_2, \ldots, k_n$, not all zero, such that

$$k_1 X_1 + k_2 X_2 + \cdots + k_n X_n = 0. \tag{4-6}$$

Consider premultiplying both sides of (4-6) by

$$(A - \lambda_2 I)(A - \lambda_3 I) \cdots (A - k_n I).$$

By use of equation (4-5), obtain

$$k_1 (A - \lambda_2 I)(A - \lambda_3 I) \cdots (A - \lambda_n I) X_1 = 0. \tag{4-7}$$

Since $(A - \lambda_1 I) X_1 = 0$, then $A X_1 = \lambda_1 X_1$. Hence, equation (4-7) may be written as

$$k_1 (\lambda_1 - \lambda_2)(\lambda_1 - \lambda_3) \cdots (\lambda_1 - \lambda_n) X_1 = 0,$$

which implies $k_1 = 0$. Similarly, it can be shown that $k_2 = k_3 = \cdots = k_n = 0$, which is contrary to the hypothesis. Therefore, the set of eigenvectors are linearly independent.

It should be noted that if the eigenvalues of a matrix are not distinct, the associated eigenvectors may or may not be linearly independent. For example, consider the matrices

$$A = \begin{pmatrix} 3 & 0 \\ 0 & 3 \end{pmatrix} \quad \text{and} \quad B = \begin{pmatrix} 3 & 1 \\ 0 & 3 \end{pmatrix}.$$

Both matrices have $\lambda_1 = \lambda_2 = 3$; that is, an eigenvalue of multiplicity two. Any nonzero vector of the form $(x_1 \quad x_2)^T$ is an eigenvector of A for λ_1 and λ_2. Hence, it is possible to choose any two linearly independent vectors such as $(1 \quad 0)^T$ and $(0 \quad 1)^T$ as eigenvectors of A that are associated with λ_1 and λ_2, respectively. Only a vector of the form $(x_1 \quad 0)^T$ is, however, an eigenvector of B for λ_1 and λ_2. Any two vectors of this form are linearly dependent; that is, one is a linear function of the other.

Theorem 4-2 *If A is a Hermitian matrix, then the eigenvalues of A are real.*

Proof: Let A be a Hermitian matrix, λ_i be any eigenvalue of A, and X_i be an eigenvector associated with λ_i. Then

$$(A - \lambda_i I)X_i = 0$$
$$AX_i - \lambda_i X_i = 0$$
$$X_i^* A X_i - \lambda_i X_i^* X_i = 0.$$

Since every eigenvector is a nonzero vector, $X_i^* X_i$ is a nonzero real number and

$$\lambda_i = \frac{X_i^* A X_i}{X_i^* X_i}.$$

Furthermore,

$$
\begin{aligned}
X_i^* A X_i &= X_i^* A^* X_i && \text{since } A = A^* \\
&= (X_i^* A X_i)^* && \text{by Example 3 of § 1-5} \\
&= \overline{X_i^* A X_i} && \text{since } X_i^* A X_i \text{ is a matrix with one element;}
\end{aligned}
$$

that is, $X_i^* A X_i$ equals its own conjugate, and hence is real. Therefore, λ_i is equal to the quotient of two real numbers, and is real.

Theorem 4-3 *If A is a real symmetric matrix, then the eigenvalues of A are real.*

Proof: Since every real symmetric matrix is a Hermitian matrix, the proof follows from Theorem 4-2.

Before presenting the next theorem it is necessary to consider the following definition: two complex eigenvectors X_1 and X_2 are defined as orthogonal if $X_1^* X_2 = 0$. For example, if $X_1 = (-i \quad 2)^T$ and $X_2 = (2i \quad 1)^T$, then $X_1^* X_2 = (i \quad 2)(2i \quad 1)^T = 0$. Hence, X_1 and X_2 are orthogonal.

Theorem 4-4 *If A is a Hermitian matrix, then the eigenvectors of A associated with distinct eigenvalues are mutually orthogonal vectors.*

Proof: Let A be a Hermitian matrix, and let X_1 and X_2 be eigenvectors associated with any two distinct eigenvalues λ_1 and λ_2, respectively. Then

$$(A - \lambda_1 I)X_1 = 0 \quad \text{and} \quad (A - \lambda_2 I)X_2 = 0;$$

that is,

$$AX_1 = \lambda_1 X_1 \quad \text{and} \quad AX_2 = \lambda_2 X_2. \tag{4-8}$$

Multiplying both sides of the first equation of (4-8) by X_2^*,

$$
\begin{aligned}
\lambda_1 X_2^* X_1 &= X_2^* A X_1 \\
&= X_2^* A^* X_1 && \text{since } A = A^* \\
&= (A X_2)^* X_1 && \text{by Example 3 of § 1-5} \\
&= \lambda_2 X_2^* X_1 && \text{by the second equation of (4-8).}
\end{aligned}
$$

Then

$$X_2^* X_1 = \frac{\lambda_2}{\lambda_1} X_2^* X_1.$$

Since λ_1 and λ_2 are real and unequal, $X_2^* X_1$ must be zero. Hence, X_1 and X_2 are orthogonal eigenvectors.

Theorem 4-5 *If A is a real symmetric matrix, then the eigenvectors of A associated with distinct eigenvalues are mutually orthogonal vectors.*

Proof: Since every real symmetric matrix is a Hermitian matrix, the proof follows from Theorem 4-4.

Exercises

1. Verify Theorem 4-1 for the matrix

$$\begin{pmatrix} 3 & 5 \\ 4 & 4 \end{pmatrix}.$$

2. Verify Theorems 4-2 and 4-4 for the matrix

$$\begin{pmatrix} 0 & 1+i \\ 1-i & 1 \end{pmatrix}.$$

3. Verify Theorems 4-3 and 4-5 for the matrix

$$\begin{pmatrix} 2 & 2 & 0 \\ 2 & 2 & 0 \\ 0 & 0 & 1 \end{pmatrix}.$$

4. Prove that if X_i is a unit eigenvector associated with the eigenvalue λ_i of A, then $X_i^T A X_i = (\lambda_i)$.

5. Prove that the eigenvalues of A^* are the conjugates of the eigenvalues of A. (*Hint:* Use the results of Exercise 4.)

4-4 Diagonalization of Matrices

It has been noted that an eigenvector X_i such that $(A - \lambda_i I)X_i = 0$, for $i = 1, 2, \ldots, n$, may be associated with each eigenvalue λ_i. This relationship may be expressed in the alternate form

$$AX_i = \lambda_i X_i, \quad \text{for} \quad i = 1, 2, \ldots, n. \tag{4-9}$$

If a square matrix of order n whose columns are eigenvectors X_i of A is constructed and denoted by X, then the equations of (4-9) may be expressed in the form

$$AX = X\Lambda, \tag{4-10}$$

where Λ is a diagonal matrix whose diagonal elements are the eigenvalues of A; that is,

$$\Lambda = \begin{pmatrix} \lambda_1 & 0 & \cdots & 0 \\ 0 & \lambda_2 & \cdots & 0 \\ \cdots & \cdots & \cdots & \cdots \\ 0 & 0 & \cdots & \lambda_n \end{pmatrix}. \tag{4-11}$$

It has been proved that the eigenvectors associated with distinct eigenvalues are linearly independent (Theorem 4-1). Hence, the matrix X will be nonsingular if the λ_i's are distinct. If both sides of equation (4-10) are multiplied by X^{-1}, the result is

$$X^{-1}AX = \Lambda. \tag{4-12}$$

Thus, by use of a matrix of eigenvectors and its inverse, it is possible to transform any matrix A with distinct eigenvalues to a diagonal matrix whose diagonal elements are the eigenvalues of A. The transformation expressed by (4-12) is referred to as the **diagonalization** of matrix A. If the eigenvalues are not distinct, the diagonalization of matrix A may not be possible. For example, the matrix

$$A = \begin{pmatrix} 3 & 1 \\ 0 & 3 \end{pmatrix}$$

cannot be diagonalized as in (4-12).

A matrix such as matrix A in equation (4-12) sometimes is spoken of as being **similar** to the diagonal matrix. In general, if there exists a nonsingular matrix C such that $C^{-1}AC = B$ for any two square matrices A and B of the same order, then A and B are called **similar matrices**, and the transformation of A to B is called a **similarity transformation**. Furthermore, if B is a diagonal matrix whose diagonal elements are the eigenvalues of A, then B is called the **classical canonical form** of matrix A. It is a unique matrix except for the order in which the eigenvalues appear along the principal diagonal.

The matrix X of (4-12) whose columns are eigenvectors of matrix A often is called a **modal matrix** of A. Recall that each eigenvector may be multiplied by any nonzero scalar. Hence, a modal matrix of A is not unique.

Example 1 Determine if

$$A = \begin{pmatrix} 6 & 2 \\ -2 & 1 \end{pmatrix} \quad \text{and} \quad B = \begin{pmatrix} 8 & 6 \\ -3 & -1 \end{pmatrix}$$

are similar matrices.

If A and B are similar matrices, there exists a nonsingular square matrix C of order two such that $C^{-1}AC = B$; that is, $AC = CB$. Let

$$C = \begin{pmatrix} a & b \\ c & d \end{pmatrix}$$

where $ad - bc \neq 0$. Then

$$\begin{pmatrix} 6 & 2 \\ -2 & 1 \end{pmatrix}\begin{pmatrix} a & b \\ c & d \end{pmatrix} = \begin{pmatrix} a & b \\ c & d \end{pmatrix}\begin{pmatrix} 8 & 6 \\ -3 & -1 \end{pmatrix}$$

$$\begin{pmatrix} 6a + 2c & 6b + 2d \\ -2a + c & -2b + d \end{pmatrix} = \begin{pmatrix} 8a - 3b & 6a - b \\ 8c - 3d & 6c - d \end{pmatrix}.$$

This single matrix equation leads to the system of homogeneous equations

$$\begin{cases} 2a - 3b - 2c & = 0 \\ 2a & + 7c - 3d = 0 \\ 6a - 7b & - 2d = 0 \\ & 2b + 6c - 2d = 0. \end{cases}$$

The system of homogeneous equations has an infinite number of solutions of the form $a = 3t - 7s$, $b = 2t - 6s$, $c = 2s$, and $d = 2t$, where s and t are arbitrary real scalars. Therefore, a nonsingular matrix

$$C = \begin{pmatrix} 3t - 7s & 2t - 6s \\ 2s & 2t \end{pmatrix}$$

exists, where s and t are arbitrary real scalars, provided $\det C \neq 0$; that is, provided

$$6t^2 - 18st + 12s^2 \neq 0,$$
$$(6t - 12s)(t - s) \neq 0,$$
$$t \neq 2s \quad \text{and} \quad t \neq s.$$

Hence, A and B are similar matrices.

As an illustration that A and B are similar matrices, let $s = 0$ and $t = 1$. Then

$$C = \begin{pmatrix} 3 & 2 \\ 0 & 2 \end{pmatrix}, \quad C^{-1} = \begin{pmatrix} \frac{1}{3} & -\frac{1}{3} \\ 0 & \frac{1}{2} \end{pmatrix},$$

and

$$C^{-1}AC = \begin{pmatrix} \frac{1}{3} & -\frac{1}{3} \\ 0 & \frac{1}{2} \end{pmatrix}\begin{pmatrix} 6 & 2 \\ -2 & 1 \end{pmatrix}\begin{pmatrix} 3 & 2 \\ 0 & 2 \end{pmatrix} = \begin{pmatrix} 8 & 6 \\ -3 & -1 \end{pmatrix} = B.$$

Example 2 Prove that similar matrices have equal determinants and equal eigenvalues.

Let A and B be similar matrices. Then a nonsingular square matrix C of the same order as A and B such that $C^{-1}AC = B$ exists. Since the determinant of the product of two matrices is equal to the product of their determinants, it follows that

$$\det B = \det C^{-1} \det A \det C$$
$$= \det C^{-1} \det C \det A$$
$$= \det (C^{-1}C) \det A$$
$$= \det I \det A$$
$$= \det A.$$

Then

$$\det (A - \lambda I) = \det [C^{-1}(A - \lambda I)C]$$
$$= \det (C^{-1}AC - \lambda C^{-1}IC)$$
$$= \det (B - \lambda I);$$

that is, A and B have the same characteristic function. An immediate consequence of this is that A and B have the same characteristic equation and eigenvalues.

Careful note should be made that the converse of the statement of Example 2 is not necessarily true. For example, the matrices

$$A = \begin{pmatrix} 1 & 0 \\ 0 & 1 \end{pmatrix} \quad \text{and} \quad B = \begin{pmatrix} 1 & 2 \\ 0 & 1 \end{pmatrix}$$

have the same eigenvalues $\lambda_1 = \lambda_2 = 1$ and $\det A = \det B$, but $C^{-1}AC = I$ for any nonsingular square matrix C of order two and $I \neq B$. Hence, A and B cannot be similar matrices.

Example 3 Prove that $A^n X_i = \lambda_i^n X_i$ for all natural numbers n.

This useful relation may be proved by mathematical induction.

$$AX_i = \lambda_i X_i \quad \text{by (4-9).}$$

Next, assume that $A^k X_i = \lambda_i^k X_i$, where k is any positive integer. Then

$$A^{k+1} X_i = A\lambda_i^k X_i$$
$$= \lambda_i^k A X_i$$
$$= \lambda_i^{k+1} X_i \quad \text{by (4-9).}$$

Hence,

$$A^n X_i = \lambda_i^n X_i \tag{4-13}$$

for all natural numbers n.

It has been shown that if A is a real symmetric matrix of order n with n distinct real eigenvalues, the associated eigenvectors are mutually orthogonal (Theorem 4-5). A matrix of eigenvectors can be made proper orthogonal if each eigenvector is normalized by an appropriate scalar multiple. A similarity transformation employing an orthogonal modal matrix is called an **orthogonal transformation**; that is, an orthogonal transformation of matrix A is of the form $C^T A C$ where C is an orthogonal matrix.

If a real symmetric matrix of order n has multiple eigenvalues, it is always possible to determine n mutually orthogonal unit eigenvectors. It is possible to show that r linearly independent eigenvectors which are orthogonal to the other eigenvectors may be associated with an eigenvalue of multiplicity r. Furthermore, it is always possible to choose these vectors orthogonal to each other. These properties of symmetric matrices will be postulated for our purposes here and their proofs left for more advanced texts in linear algebra.

Theorem 4-6 *Every real symmetric matrix can be orthogonally transformed to the classical canonical form.*

Theorem 4-6 is sometimes called the **Principal Axes Theorem.** An application of this theorem to analytic geometry will be considered in later sections of this chapter.

Example 4 Determine a proper orthogonal modal matrix which transforms the matrix A to the classical canonical form where

$$A = \begin{pmatrix} 3 & 1 \\ 1 & 3 \end{pmatrix}.$$

The characteristic equation of A is $\lambda^2 - 6\lambda + 8 = 0$; then the eigenvalues of A are $\lambda_1 = 2$ and $\lambda_2 = 4$. Associated with the eigenvalue $\lambda_1 = 2$ is the unit eigenvector

$$\left(\frac{1}{\sqrt{2}} \quad \frac{-1}{\sqrt{2}} \right)^T.$$

Associated with the eigenvalue $\lambda_2 = 4$ is the unit eigenvector

$$\left(\frac{1}{\sqrt{2}} \quad \frac{1}{\sqrt{2}} \right)^T.$$

Therefore, a proper orthogonal modal matrix which transforms the matrix A to canonical form is

$$\begin{pmatrix} \dfrac{1}{\sqrt{2}} & \dfrac{1}{\sqrt{2}} \\ \dfrac{-1}{\sqrt{2}} & \dfrac{1}{\sqrt{2}} \end{pmatrix};$$

that is,

$$\begin{pmatrix} \dfrac{1}{\sqrt{2}} & \dfrac{-1}{\sqrt{2}} \\ \dfrac{1}{\sqrt{2}} & \dfrac{1}{\sqrt{2}} \end{pmatrix} \begin{pmatrix} 3 & 1 \\ 1 & 3 \end{pmatrix} \begin{pmatrix} \dfrac{1}{\sqrt{2}} & \dfrac{1}{\sqrt{2}} \\ \dfrac{-1}{\sqrt{2}} & \dfrac{1}{\sqrt{2}} \end{pmatrix} = \begin{pmatrix} 2 & 0 \\ 0 & 4 \end{pmatrix}.$$

Exercises

In Exercises 1 and 2 determine whether or not the matrices of each pair are similar matrices.

1. $\begin{pmatrix} -2 & -1 \\ 0 & 11 \end{pmatrix}$ and $\begin{pmatrix} 0 & 1 \\ 2 & 3 \end{pmatrix}$.   **2.** $\begin{pmatrix} 2 & 2 \\ 1 & 3 \end{pmatrix}$ and $\begin{pmatrix} 2 & 1 \\ 0 & 2 \end{pmatrix}$.

3. Verify the results of Example 2 where

$$A = \begin{pmatrix} 2 & 0 \\ 1 & 1 \end{pmatrix} \quad \text{and} \quad B = \begin{pmatrix} 4 & 3 \\ -2 & -1 \end{pmatrix}.$$

4. Show that

$$A = \begin{pmatrix} 2 & 0 \\ 0 & 2 \end{pmatrix} \quad \text{and} \quad B = \begin{pmatrix} 2 & 3 \\ 0 & 2 \end{pmatrix}$$

have the same eigenvalues but are not similar matrices.

5. Verify the results of Example 3 for each eigenvalue of A where

$$A = \begin{pmatrix} 2 & 3 \\ 0 & -1 \end{pmatrix}$$

and $n = 3$.

In Exercises 6 and 7 determine a modal matrix which transforms the matrix to the classical canonical form. Perform the transformation. Check the results.

6. $\begin{pmatrix} 2 & 1 \\ 0 & 3 \end{pmatrix}$.   **7.** $\begin{pmatrix} 5 & 4 \\ 12 & 7 \end{pmatrix}$.

In Exercises 8 and 9 determine a proper orthogonal modal matrix which transforms the matrix to the classical canonical form. Perform the orthogonal transformation. Check the results.

8. $\begin{pmatrix} 4 & 6 \\ 6 & -1 \end{pmatrix}$.   **9.** $\begin{pmatrix} 2 & 2 & 0 \\ 2 & 2 & 0 \\ 0 & 0 & 1 \end{pmatrix}$.

10. Prove that any matrix A similar to a diagonal matrix is similar to A^T.

11. Prove that if A is similar to the scalar matrix kI, then $A = kI$.

4-5 The Hamilton-Cayley Theorem

An important and interesting theorem of the theory of matrices is the **Hamilton-Cayley Theorem**:

Theorem 4-7 *Every square matrix A satisfies its own characteristic equation* $|A - \lambda I| = 0$.

More precisely, if λ is replaced by the matrix A of order n and each real number c_n is replaced by the scalar multiple $c_n I$ where I is the identity matrix of order n, then the characteristic equation of matrix A becomes a valid matrix equation; that is,

$$c_0 A^n + c_1 A^{n-1} + \cdots + c_{n-1} A + c_n I = 0. \qquad (4\text{-}14)$$

A heuristic argument may be used to prove the Hamilton-Cayley Theorem for a matrix A with distinct eigenvalues. Replace the variable λ by the square matrix A and c_n by $c_n I$ in the expression for the characteristic function of A and obtain

$$f(A) = c_0 A^n + c_1 A^{n-1} + \cdots + c_{n-1} A + c_n I. \qquad (4\text{-}15)$$

Postmultiply both sides of equation (4-15) by an eigenvector X_i of A associated with λ_i and obtain

$$f(A)X_i = (c_0 \lambda_i^n + c_1 \lambda^{n-1} + \cdots + c_{n-1}\lambda_i + c_n)X_i$$

since $A^k X_i = \lambda_i^k X_i$ by Example 3 of §4-4. Since

$$c_0 \lambda_i^n + c_1 \lambda_i^{n-1} + \cdots + c_{n-1}\lambda_i + c_n = 0 \quad \text{for} \quad i = 1, 2, \ldots, n,$$

then

$$f(A)X_i = 0 \quad \text{for} \quad i = 1, 2, \ldots, n.$$

Hence,

$$f(A)X = 0, \qquad (4\text{-}16)$$

where X is a matrix of eigenvectors. Since the eigenvectors are linearly independent by Theorem 4-1, the matrix of eigenvectors has a unique inverse X^{-1}. If both sides of equation (4-16) are postmultiplied by X^{-1}, the result is $f(A) = 0$, and the theorem is proved.

Proofs of the Hamilton-Cayley Theorem for the general case without restrictions on the eigenvalues of A may be found in most advanced texts on linear algebra.

Example 1 Show that the matrix A where

$$A = \begin{pmatrix} 3 & -2 \\ 1 & 2 \end{pmatrix}$$

satisfies its own characteristic equation.

The characteristic function of A is

$$f(\lambda) = |A - \lambda I| = \begin{vmatrix} 3 - \lambda & -2 \\ 1 & 2 - \lambda \end{vmatrix};$$

that is, $f(\lambda) = \lambda^2 - 5\lambda + 8$. Replace λ by A and 8 by $8I$ where I is the identity matrix of order two and obtain

$$f(A) = \begin{pmatrix} 3 & -2 \\ 1 & 2 \end{pmatrix}^2 - 5\begin{pmatrix} 3 & -2 \\ 1 & 2 \end{pmatrix} + 8\begin{pmatrix} 1 & 0 \\ 0 & 1 \end{pmatrix}$$

$$= \begin{pmatrix} 7 & -10 \\ 5 & 2 \end{pmatrix} + \begin{pmatrix} -15 & 10 \\ -5 & -10 \end{pmatrix} + \begin{pmatrix} 8 & 0 \\ 0 & 8 \end{pmatrix} = \begin{pmatrix} 0 & 0 \\ 0 & 0 \end{pmatrix}.$$

Hence, $f(A) = 0$ and the Hamilton-Cayley Theorem has been verified for matrix A.

The Hamilton-Cayley Theorem may be applied to the problem of determining the inverse of a nonsingular matrix A. Let

$$c_0\lambda^n + c_1\lambda^{n-1} + \cdots + c_{n-1}\lambda + c_n = 0$$

be the characteristic equation of A. Note that since A is a nonsingular matrix, $\lambda_i \neq 0$; that is, every eigenvalue is nonzero, and $c_n \neq 0$. By the Hamilton-Cayley Theorem,

$$c_0A^n + c_1A^{n-1} + \cdots + c_{n-1}A + c_nI = 0,$$

and

$$I = -\frac{1}{c_n}(c_0A^n + c_1A^{n-1} + \cdots + c_{n-1}A). \tag{4-17}$$

If both sides of (4-17) are multiplied by A^{-1}, the result is

$$A^{-1} = -\frac{1}{c_n}(c_0A^{n-1} + c_1A^{n-2} + \cdots + c_{n-1}I). \tag{4-18}$$

Note that the calculation of an inverse by use of equation (4-18) is quite adaptable to high-speed digital computers and is not difficult to compute manually for small values of n. In calculating the powers of matrix A necessary in equation (4-18), the necessary information concerning $tr(A^k)$ for calculating the c_i's is also obtained.

Example 2 Use the Hamilton-Cayley Theorem to find the inverse of A where

$$A = \begin{pmatrix} 1 & 0 & 1 \\ -1 & 1 & -3 \\ 2 & 2 & 4 \end{pmatrix}.$$

The characteristic equation of A is

$$\begin{vmatrix} 1-\lambda & 0 & 1 \\ -1 & 1-\lambda & -3 \\ 2 & 2 & 4-\lambda \end{vmatrix} = 0;$$

that is, $\lambda^3 - 6\lambda^2 + 13\lambda - 6 = 0$. By the Hamilton-Cayley Theorem,

$$A^3 - 6A^2 + 13A - 6I = 0,$$

$$I = \tfrac{1}{6}(A^3 - 6A^2 + 13A),$$

and
$$A^{-1} = \tfrac{1}{6}(A^2 - 6A + 13I).$$
Therefore,

$$A^{-1} = \tfrac{1}{6}\left[\begin{pmatrix} 1 & 0 & 1 \\ -1 & 1 & -3 \\ 2 & 2 & 4 \end{pmatrix}^2 - 6\begin{pmatrix} 1 & 0 & 1 \\ -1 & 1 & -3 \\ 2 & 2 & 4 \end{pmatrix} + 13\begin{pmatrix} 1 & 0 & 0 \\ 0 & 1 & 0 \\ 0 & 0 & 1 \end{pmatrix}\right]$$

$$= \tfrac{1}{6}\left[\begin{pmatrix} 3 & 2 & 5 \\ -8 & -5 & -16 \\ 8 & 10 & 12 \end{pmatrix} + \begin{pmatrix} -6 & 0 & -6 \\ 6 & -6 & 18 \\ -12 & -12 & -24 \end{pmatrix} + \begin{pmatrix} 13 & 0 & 0 \\ 0 & 13 & 0 \\ 0 & 0 & 13 \end{pmatrix}\right]$$

$$= \begin{pmatrix} \tfrac{5}{3} & \tfrac{1}{3} & -\tfrac{1}{6} \\ -\tfrac{1}{3} & \tfrac{1}{3} & \tfrac{1}{3} \\ -\tfrac{2}{3} & -\tfrac{1}{3} & \tfrac{1}{6} \end{pmatrix}.$$

A^{-k}, where k is a positive integer, is defined to be equal to $(A^{-1})^k$. By use of equation (4-18), it is now possible to express any negative integral power of a nonsingular matrix A of order n in terms of a linear function of the first $(n - 1)$ powers of A.

Example 3 Find A^{-3} for the nonsingular matrix A where
$$A = \begin{pmatrix} 2 & 4 \\ 1 & 1 \end{pmatrix}.$$
Verify that A^{-3} is the inverse of A^3.

The characteristic equation of A is $\lambda^2 - 3\lambda - 2 = 0$, and $c_0 = 1$, $c_1 = -3$, and $c_2 = -2$. By use of equation (4-18),
$$A^{-1} = \tfrac{1}{2}(A - 3I).$$
Then
$$A^{-2} = A^{-1}A^{-1} = \tfrac{1}{2}(I - 3A^{-1}) = \tfrac{11}{4}I - \tfrac{3}{4}A,$$
and
$$A^{-3} = A^{-1}A^{-2} = \tfrac{11}{4}A^{-1} - \tfrac{3}{4}I = \tfrac{11}{8}A - \tfrac{39}{8}I.$$
Therefore,
$$A^{-3} = \tfrac{11}{8}\begin{pmatrix} 2 & 4 \\ 1 & 1 \end{pmatrix} - \tfrac{39}{8}\begin{pmatrix} 1 & 0 \\ 0 & 1 \end{pmatrix} = \begin{pmatrix} -\tfrac{17}{8} & \tfrac{11}{2} \\ \tfrac{11}{8} & -\tfrac{7}{2} \end{pmatrix}.$$
Verify this result by noting that
$$A^3 = \begin{pmatrix} 28 & 44 \\ 11 & 17 \end{pmatrix}, \quad \text{and} \quad \begin{pmatrix} 28 & 44 \\ 11 & 17 \end{pmatrix}\begin{pmatrix} -\tfrac{17}{8} & \tfrac{11}{2} \\ \tfrac{11}{8} & -\tfrac{7}{2} \end{pmatrix} = \begin{pmatrix} 1 & 0 \\ 0 & 1 \end{pmatrix}.$$

Exercises

In Exercises 1 through 4 verify the Hamilton-Cayley Theorem for the given matrix A.

1. $\begin{pmatrix} 2 & -1 \\ 3 & 4 \end{pmatrix}$.

2. $\begin{pmatrix} 7 & 9 \\ 2 & 1 \end{pmatrix}$.

3. $\begin{pmatrix} 5 & 0 \\ 0 & 2 \end{pmatrix}$.

4. $\begin{pmatrix} 1 & 1 & -2 \\ 1 & 0 & 3 \\ -2 & 3 & 2 \end{pmatrix}$.

In Exercises 5 and 6 use the Hamilton-Cayley Theorem to find the inverse of the given matrix A.

5. $\begin{pmatrix} 5 & 2 \\ 2 & 1 \end{pmatrix}$.

6. $\begin{pmatrix} 1 & 2 & 3 \\ 1 & 3 & 5 \\ 1 & 5 & 12 \end{pmatrix}$.

7. Find (a) A^{-2}; (b) A^{-3} for the nonsingular matrix A where

$$A = \begin{pmatrix} 7 & 2 \\ 5 & 1 \end{pmatrix}.$$

Verify that A^{-2} and A^{-3} are the inverses of A^2 and A^3, respectively.

8. Prove the Hamilton-Cayley Theorem for any square matrix of order two.

4-6 Quadratic Forms

Prior to a discussion of the simplification of the general quadratic equation in two variables by means of the rigid motion transformations, it is necessary to introduce the concept of real quadratic forms. A **quadratic form** is a homogeneous polynomial of the second degree in n variables $w_1, w_2, \ldots, w_n$; that is, a polynomial of the form

$$\sum_{i, j=1}^{n} a_{ij} w_i w_j. \tag{4-19}$$

Every quadratic form may be expressed as a matrix product $W^T A W$, where $W = (w_1 \quad w_2 \quad \cdots \quad w_n)^T$ and $A = ((a_{ij}))$ is a unique real symmetric matrix. Therefore, every real quadratic form may be orthogonally transformed to the form $\lambda_1 z_1^2 + \lambda_2 z_2^2 + \cdots + \lambda_n z_n^2$, where the λ_i's are the eigenvalues of A, by an appropriate choice of a proper orthogonal matrix C such that $C^T A C$ is a diagonal matrix; that is,

$$W^T A W = Z^T C^T A C Z = Z^T \Lambda Z, \tag{4-20}$$

where $W = CZ$ and $Z = (z_1 \quad z_2 \quad \cdots \quad z_n)^T$. Of course, the appropriate orthogonal matrix C which transforms a real symmetric matrix A into a diagonal matrix is a matrix of unit eigenvectors of A.

If (4-19) is a homogeneous polynomial of the second degree in two variables, the modal matrix C of the orthogonal transformation in (4-20) may be considered a rotation matrix. For example, the problem of reducing the quadratic form $ax^2 + 2bxy + cy^2$ to the canonical form $a'x'^2 + c'y'^2$ can be accomplished by means of a rotation of the plane about the origin. When the quadratic form is written in matrix notation, the result is

$$(x \quad y)\begin{pmatrix} a & b \\ b & c \end{pmatrix}\begin{pmatrix} x \\ y \end{pmatrix}.$$

A rotation of the plane about the origin through an angle θ may be expressed by the matrix equation

$$\begin{pmatrix} x \\ y \end{pmatrix} = \begin{pmatrix} \cos\theta & \sin\theta \\ -\sin\theta & \cos\theta \end{pmatrix}\begin{pmatrix} x' \\ y' \end{pmatrix}.$$

Since the transpose of a product of two matrices is equal to the product of the transpose of the two matrices taken in reverse order, then

$$(x' \quad y')\begin{pmatrix} \cos\theta & -\sin\theta \\ \sin\theta & \cos\theta \end{pmatrix}\begin{pmatrix} a & b \\ b & c \end{pmatrix}\begin{pmatrix} \cos\theta & \sin\theta \\ -\sin\theta & \cos\theta \end{pmatrix}\begin{pmatrix} x' \\ y' \end{pmatrix}$$

represents the quadratic form after a rotation of the plane about the origin through an angle θ. It is desired that θ, the angle of rotation, be such that

$$\begin{pmatrix} \cos\theta & -\sin\theta \\ \sin\theta & \cos\theta \end{pmatrix}\begin{pmatrix} a & b \\ b & c \end{pmatrix}\begin{pmatrix} \cos\theta & \sin\theta \\ -\sin\theta & \cos\theta \end{pmatrix} = \begin{pmatrix} a' & 0 \\ 0 & c' \end{pmatrix}.$$

This is clearly the problem of transforming a symmetric matrix to a diagonal matrix. Previously, it has been shown that in such a case the diagonal elements of the diagonal matrix are the eigenvalues of the symmetric matrix. Therefore, to accomplish the proper rotation, it is necessary only to solve the characteristic equation of the quadratic form:

$$\begin{vmatrix} a - \lambda & b \\ b & c - \lambda \end{vmatrix} = 0;$$

that is,

$$\lambda^2 - (a + c)\lambda + (ac - b^2) = 0.$$

The eigenvalues λ_1 and λ_2 become the coefficients of x'^2 and y'^2; that is,

$$(x \quad y)\begin{pmatrix} a & b \\ b & c \end{pmatrix}\begin{pmatrix} x \\ y \end{pmatrix} = (x' \quad y')\begin{pmatrix} \lambda_1 & 0 \\ 0 & \lambda_2 \end{pmatrix}\begin{pmatrix} x' \\ y' \end{pmatrix} = (\lambda_1 x'^2 + \lambda_2 y'^2).$$

The angle of rotation θ may be determined by considering the matrix equation

$$\begin{pmatrix} \cos\theta & -\sin\theta \\ \sin\theta & \cos\theta \end{pmatrix} \begin{pmatrix} a & b \\ b & c \end{pmatrix} \begin{pmatrix} \cos\theta & \sin\theta \\ -\sin\theta & \cos\theta \end{pmatrix} = \begin{pmatrix} \lambda_1 & 0 \\ 0 & \lambda_2 \end{pmatrix};$$

that is,

$$\begin{cases} a\cos^2\theta - b\sin\theta\cos\theta - b\sin\theta\cos\theta + c\sin^2\theta = \lambda_1 \\ a\sin\theta\cos\theta + b\cos^2\theta - b\sin^2\theta - c\sin\theta\cos\theta = 0 \\ a\sin\theta\cos\theta - b\sin^2\theta + b\cos^2\theta - c\sin\theta\cos\theta = 0 \\ a\sin^2\theta + b\sin\theta\cos\theta + b\sin\theta\cos\theta + c\cos^2\theta = \lambda_2. \end{cases} \qquad (4\text{-}21)$$

If the second and third equations of (4-21) are added, the result is

$$(a - c)\sin\theta\cos\theta + b(\cos^2\theta - \sin^2\theta) = 0.$$

Therefore,

$$\frac{2b}{c - a} = \frac{\sin 2\theta}{\cos 2\theta} = \tan 2\theta.$$

Hence,

$$\theta = \tfrac{1}{2}\arctan\frac{2b}{c - a}. \qquad (4\text{-}22)$$

Note that the unit vectors along the coordinate axes are the images of the unit eigenvectors $(\cos\theta \quad -\sin\theta)^T$ and $(\sin\theta \quad \cos\theta)^T$ associated with the eigenvalues of the symmetric matrix of the quadratic form. These unit eigenvectors lie along the principal axes of the locus of $ax^2 + 2bxy + y^2 = 0$.

Exercises

In Exercises 1 through 3 express each quadratic form in terms of matrices.

1. $3x^2 + 10xy + 3y^2$.   **2.** $x^2 - 2xy + y^2$.   **3.** $2x^2 + 2\sqrt{2}xy + y^2$.

In Exercises 4 through 6 use a proper orthogonal matrix to transform each quadratic form in the specified exercise to canonical form.

4. Exercise 1.   **5.** Exercise 2.   **6.** Exercise 3.

4-7 Classification of the Conics

The most general equation of the second degree in two variables x and y may be written in the form

$$f(x, y) = ax^2 + 2bxy + cy^2 + 2dx + 2ey + f = 0, \qquad (4\text{-}23)$$

where $a, b, c, d, e,$ and f are real numbers. The locus of equation (4-23) on the coordinate Euclidean plane is called a **conic section**, or simply a **conic**. By means of a rotation of the plane about the origin, a translation of the plane, or both, it is possible to represent every conic in a simplified standard, or canonical, form. These plane figures can be studied more readily in their canonical forms. There are nine classes of the conics; that is, equation (4-23) represents one of nine types of plane figures called conics. Two of these plane figures are imaginary in that there are no real points which satisfy equation (4-23).

If the general second degree function $f(x, y)$ of equation (4-23) is written in the form

$$
\begin{aligned}
f(x, y) = \; & axx + bxy + dx \\
& + byx + cyy + ey \\
& + dx + ey + f,
\end{aligned}
\tag{4-24}
$$

another valuable form of $f(x, y)$ is suggested. The function may be expressed as the product of three matrices:

$$
(x \quad y \quad 1)\begin{pmatrix} a & b & d \\ b & c & e \\ d & e & f \end{pmatrix}\begin{pmatrix} x \\ y \\ 1 \end{pmatrix}.
\tag{4-25}
$$

Note that the matrix of central importance

$$
\Delta = \begin{pmatrix} a & b & d \\ b & c & e \\ d & e & f \end{pmatrix},
\tag{4-26}
$$

which defines the particular conic being studied, is a real symmetric matrix. The matrix Δ is called the **matrix of the conic section.**

In addition to Δ, the matrix whose determinant is the minor of f in Δ, the real symmetric matrix

$$
F = \begin{pmatrix} a & b \\ b & c \end{pmatrix},
\tag{4-27}
$$

is of fundamental importance in the analysis of $f(x, y)$. The result of a pre-multiplication of matrix F by row vector $(x \quad y)$ and a postmultiplication of matrix F by column vector $(x \quad y)^T$ represents the quadratic form portion of the general second degree function; that is,

$$
(x \quad y)\begin{pmatrix} a & b \\ b & c \end{pmatrix}\begin{pmatrix} x \\ y \end{pmatrix} = ax^2 + 2bxy + cy^2.
$$

After a rotation of the plane about the origin through an angle θ defined by the matrix equation

$$
\begin{pmatrix} x \\ y \\ 1 \end{pmatrix} = \begin{pmatrix} \cos\theta & \sin\theta & 0 \\ -\sin\theta & \cos\theta & 0 \\ 0 & 0 & 1 \end{pmatrix}\begin{pmatrix} x' \\ y' \\ 1 \end{pmatrix},
$$

where θ is $\frac{1}{2}$ arc tan $2b/(c - a)$, the general equation of a conic may be expressed in the form

$$\lambda_1 x'^2 + \lambda_2 y'^2 + 2\alpha x' + 2\beta y' + f = 0; \tag{4-28}$$

that is,

$$(x' \ \ y' \ \ 1) \begin{pmatrix} \lambda_1 & 0 & \alpha \\ 0 & \lambda_2 & \beta \\ \alpha & \beta & f \end{pmatrix} \begin{pmatrix} x' \\ y' \\ 1 \end{pmatrix} = 0, \tag{4-29}$$

where the λ_i's are the eigenvalues of matrix F and

$$\begin{cases} \alpha = d \cos \theta - e \sin \theta \\ \beta = d \sin \theta + e \cos \theta. \end{cases} \tag{4-30}$$

The linear terms in x' and y' of (4-28) may be removed by a translation of the plane defined by the matrix equation

$$\begin{pmatrix} x' \\ y' \\ 1 \end{pmatrix} = \begin{pmatrix} 1 & 0 & -\dfrac{\alpha}{\lambda_1} \\ 0 & 1 & -\dfrac{\beta}{\lambda_2} \\ 0 & 0 & 1 \end{pmatrix} \begin{pmatrix} x'' \\ y'' \\ 1 \end{pmatrix}. \tag{4-31}$$

If at least one λ_i equals zero, a translation of the plane which would remove all the linear terms of equation (4-28) does not exist. Hence, a geometric center for the conic does not exist and the conic is called a **noncentral conic**. If some λ_i equals zero, matrix F is necessarily singular. When F is non-singular, a geometric center for the conic exists and the conic is called a **central conic**. After a translation of the plane described by (4-31), the equation of a central conic may be expressed as

$$\lambda_1 x''^2 + \lambda_2 y''^2 + f' = 0; \tag{4-32}$$

that is,

$$(x'' \ \ y'' \ \ 1) \begin{pmatrix} \lambda_1 & 0 & 0 \\ 0 & \lambda_2 & 0 \\ 0 & 0 & f' \end{pmatrix} \begin{pmatrix} x'' \\ y'' \\ 1 \end{pmatrix} = 0, \tag{4-33}$$

where

$$f' = f - \frac{\alpha^2}{\lambda_1} - \frac{\beta^2}{\lambda_2}. \tag{4-34}$$

If one and only one λ_1 is zero, then a translation of the plane exists which would remove one of the linear terms of equation (4-28). For example, consider $\lambda_2 = 0$. The translation of the plane defined by the matrix equation

$$\begin{pmatrix} x' \\ y' \\ 1 \end{pmatrix} = \begin{pmatrix} 1 & 0 & -\dfrac{\alpha}{\lambda_1} \\ 0 & 1 & 0 \\ 0 & 0 & 1 \end{pmatrix} \begin{pmatrix} x'' \\ y'' \\ 1 \end{pmatrix} \tag{4-35}$$

transforms equation (4-28) to the form

$$\lambda_1 x''^2 + 2\beta y'' + f' = 0; \tag{4-36}$$

that is,

$$(x'' \quad y'' \quad 1)\begin{pmatrix} \lambda_1 & 0 & 0 \\ 0 & 0 & \beta \\ 0 & \beta & f' \end{pmatrix}\begin{pmatrix} x'' \\ y'' \\ 1 \end{pmatrix} = 0, \tag{4-37}$$

where

$$f' = f - \frac{\alpha^2}{\lambda_1}. \tag{4-38}$$

Example 1 Transform the equation of the conic $5x^2 + 6xy + 5y^2 - 4x + 4y - 4 = 0$ to canonical form (Figure 4-2).

The equation of the conic may be written in the form

$$(x \quad y \quad 1)\begin{pmatrix} 5 & 3 & -2 \\ 3 & 5 & 2 \\ -2 & 2 & -4 \end{pmatrix}\begin{pmatrix} x \\ y \\ 1 \end{pmatrix} = 0.$$

The characteristic equation of

$$F = \begin{pmatrix} 5 & 3 \\ 3 & 5 \end{pmatrix}$$

is $\lambda^2 - 10\lambda + 16 = 0$. Hence, the eigenvalues of F are $\lambda_1 = 8$ and $\lambda_2 = 2$ with associated unit eigenvectors

$$\left(\frac{1}{\sqrt{2}} \quad \frac{1}{\sqrt{2}}\right)^T \quad \text{and} \quad \left(-\frac{1}{\sqrt{2}} \quad \frac{1}{\sqrt{2}}\right)^T,$$

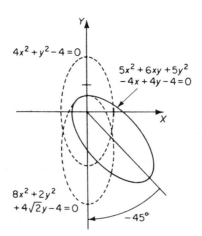

Figure 4-2

respectively. Therefore, a rotation of the plane defined by the matrix equation

$$\begin{pmatrix} x \\ y \\ 1 \end{pmatrix} = \begin{pmatrix} \dfrac{1}{\sqrt{2}} & -\dfrac{1}{\sqrt{2}} & 0 \\ \dfrac{1}{\sqrt{2}} & \dfrac{1}{\sqrt{2}} & 0 \\ 0 & 0 & 1 \end{pmatrix} \begin{pmatrix} x' \\ y' \\ 1 \end{pmatrix}$$

will transform the equation of the conic to the form of (4-28):

$$(x' \quad y' \quad 1) \begin{pmatrix} \dfrac{1}{\sqrt{2}} & \dfrac{1}{\sqrt{2}} & 0 \\ -\dfrac{1}{\sqrt{2}} & \dfrac{1}{\sqrt{2}} & 0 \\ 0 & 0 & 1 \end{pmatrix} \begin{pmatrix} 5 & 3 & -2 \\ 3 & 5 & 2 \\ -2 & 2 & -4 \end{pmatrix}$$

$$\begin{pmatrix} \dfrac{1}{\sqrt{2}} & -\dfrac{1}{\sqrt{2}} & 0 \\ \dfrac{1}{\sqrt{2}} & \dfrac{1}{\sqrt{2}} & 0 \\ 0 & 0 & 1 \end{pmatrix} \begin{pmatrix} x' \\ y' \\ 1 \end{pmatrix} = 0,$$

$$(x' \quad y' \quad 1) \begin{pmatrix} 8 & 0 & 0 \\ 0 & 2 & 2\sqrt{2} \\ 0 & 2\sqrt{2} & -4 \end{pmatrix} \begin{pmatrix} x' \\ y' \\ 1 \end{pmatrix} = 0;$$

that is,

$$8x'^2 + 2y'^2 + 4\sqrt{2}\,y' - 4 = 0.$$

A translation of the plane defined by (4-31) where $\alpha = 0$, $\beta = 2\sqrt{2}$, $\lambda_1 = 8$, and $\lambda_2 = 2$ will transform the equation of the conic to the form of (4-32):

$$(x'' \quad y'' \quad 1) \begin{pmatrix} 1 & 0 & 0 \\ 0 & 1 & 0 \\ 0 & -\sqrt{2} & 1 \end{pmatrix} \begin{pmatrix} 8 & 0 & 0 \\ 0 & 2 & 2\sqrt{2} \\ 0 & 2\sqrt{2} & -4 \end{pmatrix}$$

$$\begin{pmatrix} 1 & 0 & 0 \\ 0 & 1 & -\sqrt{2} \\ 0 & 0 & 1 \end{pmatrix} \begin{pmatrix} x'' \\ y'' \\ 1 \end{pmatrix} = 0,$$

$$(x'' \quad y'' \quad 1) \begin{pmatrix} 8 & 0 & 0 \\ 0 & 2 & 0 \\ 0 & 0 & -8 \end{pmatrix} \begin{pmatrix} x'' \\ y'' \\ 1 \end{pmatrix} = 0,$$

$$8x''^2 + 2y''^2 - 8 = 0,$$

$$4x''^2 + y''^2 - 4 = 0;$$

that is,

$$4x^2 + y^2 - 4 = 0.$$

Hence, the conic is a real ellipse. Note that the unit eigenvectors are parallel to the principal axes of the original conic.

It should be evident from the discussion thus far that the class of conic represented by equation (4-23) can be determined by an investigation of the algebraic values of the eigenvalues of matrix F. The conics will now be classified according to the eigenvalues of F.

If λ_1 and λ_2 are nonzero and have the same algebraic sign, then equation (4-32) represents an imaginary ellipse, a real ellipse, or a real point depending upon whether f' has the same sign as the λ_i's, differs in sign from the λ_i's, or is zero, respectively. Furthermore, if the eigenvalues are equal in the case of the real ellipse, the conic is a circle.

If λ_1 and λ_2 are nonzero and have opposite algebraic signs, then equation (4-32) represents a hyperbola or two intersecting lines depending upon whether f' is nonzero or zero, respectively.

For the case when only one eigenvalue is zero consider λ_2 equal to zero; there is no loss in generality in doing so. Then equation (4-23) may be transformed to the form of (4-36). If β is nonzero, then equation (4-36) represents a parabola; if β is equal to zero, then equation (4-36) represents two parallel lines if λ_1 and f' have opposite signs, two imaginary parallel lines if λ_1 and f' have the same signs, and two coincident lines if f' is equal to zero.

Example 2 Identify the class of conic represented by the equation $x^2 - 4xy + 4y^2 - 4 = 0.$

The matrix of the conic section is

$$\begin{pmatrix} 1 & -2 & 0 \\ -2 & 4 & 0 \\ 0 & 0 & -4 \end{pmatrix}.$$

The characteristic equation of matrix F is

$$\begin{vmatrix} 1 - \lambda & -2 \\ -2 & 4 - \lambda \end{vmatrix} = 0;$$

that is, $\lambda^2 - 5\lambda = 0$. Hence, the eigenvalues of F are $\lambda_1 = 5$ and $\lambda_2 = 0$ with associated unit eigenvectors

$$\left(\frac{1}{\sqrt{5}} \quad -\frac{2}{\sqrt{5}} \right)^T \quad \text{and} \quad \left(\frac{2}{\sqrt{5}} \quad \frac{1}{\sqrt{5}} \right)^T.$$

Since the inverse of the rotation matrix and the proper orthogonal matrix of eigenvectors are identical,

$$\begin{pmatrix} \cos\theta & \sin\theta \\ -\sin\theta & \cos\theta \end{pmatrix} = \begin{pmatrix} \dfrac{1}{\sqrt{5}} & \dfrac{2}{\sqrt{5}} \\ -\dfrac{2}{\sqrt{5}} & \dfrac{1}{\sqrt{5}} \end{pmatrix},$$

$\cos\theta = 1/\sqrt{5}$, and $\sin\theta = 2/\sqrt{5}$. Furthermore, $d = e = 0$; $f = -4$. Then $\alpha = \beta = 0$ by (4-30), and $f' = -4$ by (4-38). Since one eigenvalue λ_2 is zero and the remaining eigenvalue λ_1 and f' have opposite signs, the equation of the conic represents two parallel lines.

Note that the equation of the conic may be expressed in factored form as $(x - 2y + 2)(x - 2y - 2) = 0$. Hence, the equations of the two parallel lines are $x - 2y + 2 = 0$ and $x - 2y - 2 = 0$.

Exercises

In Exercises 1 through 4 transform the equation of the conic to canonical form and identify the conic.

1. $x^2 + xy + y^2 + 2x - 3y + 5 = 0$.

2. $x^2 - 2xy + y^2 + 8x + 8y = 0$.

3. $7x^2 - 48xy - 7y^2 + 60x - 170y + 225 = 0$.

4. $x^2 - 2xy + y^2 - 8 = 0$.

In Exercises 5 through 8 identify the class of conic represented by the equation.

5. $5x^2 + 4xy + 8y^2 - 36 = 0$.

6. $2x^2 + \sqrt{3}\,xy + y^2 + 14 = 0$.

7. $x^2 - 12xy - 4y^2 = 0$.

8. $3x^2 + 3xy + 3y^2 - 18x + 15y + 91 = 0$.

9. Show that the eigenvalues of matrix F in Example 1 are inversely proportional to the squares of the semi-axes.

10. Show that the geometric center (x, y) of the conic in Example 1 satisfies the system of equations

$$\begin{cases} ax + by + d = 0 \\ bx + cy + e = 0. \end{cases}$$

4-8 Invariants for Conics

The rank of the matrix of a conic section is called the **rank of the conic**. Since the rank of a matrix is an invariant under the elementary row transformations and since every rotation matrix and translation matrix may be expressed as a product of elementary row transformations, the rank of Δ is an invariant under the rigid motion transformations represented by these matrices.

If λ_1 and λ_2 are nonzero, then the transformed matrix of the conic section is represented in equation (4-33) by

$$\Delta_1 = \begin{pmatrix} \lambda_1 & 0 & 0 \\ 0 & \lambda_2 & 0 \\ 0 & 0 & f' \end{pmatrix}.$$

If f' is nonzero, Δ_1 is of rank three. Hence, from previous discussion of the classification of the conics, a real ellipse, an imaginary ellipse, and a hyperbola are conics of rank three. If f' is equal to zero, Δ_1 is of rank two. Hence, a real point and a pair of intersecting lines are conics of rank two.

If one and only one λ_i is zero (consider $\lambda_2 = 0$), then the transformed matrix of the conic section is represented in equation (4-37) by

$$\Delta_2 = \begin{pmatrix} \lambda_1 & 0 & 0 \\ 0 & 0 & \beta \\ 0 & \beta & f' \end{pmatrix}.$$

If β is nonzero, Δ_2 is of rank three. Hence, a parabola is a conic of rank three. If β is equal to zero, Δ_2 is of rank one or two depending upon whether f' is zero or nonzero, respectively. Hence, a pair of coincident lines is a conic of rank one; a pair of real parallel lines and a pair of imaginary parallel lines are conics of rank two.

The conics of rank three sometimes are called the **proper conics**; that is, the ellipses, the hyperbola, and the parabola are proper conics. The conics of rank less than three sometimes are called the **degenerate conics.**

Since a rotation of the plane about the origin transforms matrix F to a similar matrix, then by Example 2 of §4-4 the characteristic function of F is an invariant under the rotation transformation. Furthermore, since the elements of F are the coefficients of the second degree terms of $f(x, y)$ in (4-23) and since these coefficients are unchanged by a translation·of the plane, the characteristic function of F is an invariant under the translation transformation. The characteristic function of F is.

$$\lambda_2 - (a + c)\lambda + (ac - b^2). \tag{4-39}$$

Therefore, the following quantities are invariants for conics under the rotation and translation transformations:

$$a + c; \tag{4-40}$$

$$ac - b^2. \tag{4-41}$$

Example 1 Find the rank of the conic in Example 2 of § 4-7.

The matrix of the conic section in Example 2 of § 4-7 is

$$\Delta = \begin{pmatrix} 1 & -2 & 0 \\ -2 & 4 & 0 \\ 0 & 0 & -4 \end{pmatrix}.$$

Since $|\Delta| = 0$, the rank of Δ cannot be three. The rank of Δ is two since

$$\begin{vmatrix} 4 & 0 \\ 0 & -4 \end{vmatrix} \neq 0.$$

Hence, the rank of the conic is two.

Example 2 Verify the invariance of (a) $a + c$; (b) $ac - b^2$ for the conic of Example 1 of § 4-7 under the rotation and translation transformations.

In Example 1 of § 4-7, $a = 5$, $b = 3$, and $c = 5$. After the rotation and translation transformations, the coefficients of x^2, $2xy$, and y^2 are 8, 0, and 2, respectively. Therefore,

(a) $a + c = 5 + 5 = 8 + 2$;
(b) $ac - b^2 = (5)(5) - (3)^2 = (8)(2) - (0)^2$.

Exercises

Find the rank of each conic in the specified exercises of § 4-7.

1. Exercise 1.   2. Exercise 2.

3. Exercise 3.   4. Exercise 4.

Verify the invariance of (a) $a + c$; (b) $ac - b^2$ for each conic in the specified exercises of § 4-7.

5. Exercise 1.   6. Exercise 2.

7. Exercise 3.   8. Exercise 4.

9. Show that (a) $a + c = \lambda_1 + \lambda_2$; (b) $ac - b^2 = \lambda_1\lambda_2$; (c) $|F| = ac - b^2$.

10. Discuss the classification of the conics in terms of $|\Delta|$ and $|F|$. (See Exercise 9.)

11. Prove that $a + c + f$ is an invariant for conics under the rotation transformation.

12. Prove that $d^2 + e^2$ of (4-23) is an invariant for conics under the rotation transformation.

Bibliography

Andree, Richard V., *Selections from Modern Abstract Algebra*. New York: Holt, Rinehart & Winston, Inc., 1958.

Beaumont, Ross A., and Richard W. Ball, *Introduction to Modern Algebra and Matrix Theory*. New York: Holt, Rinehart & Winston, Inc., 1954.

Birkhoff, Garrett, and Saunders MacLane, *A Survey of Modern Algebra*. (3rd ed.). New York: The Macmillan Company, 1965.

Finkbeiner, Daniel T., *Introduction to Matrices and Linear Transformations*. San Francisco: W. H. Freeman & Co., Publishers, 1960.

Hoffman, Kenneth, and Ray Kunze, *Linear Algebra*. Englewood Cliffs, N.J.: Prentice-Hall, Inc., 1961.

Hohn, Franz E., *Elementary Matrix Algebra*. New York: The Macmillan Company, 1958.

Johnson, Richard E., *First Course in Abstract Algebra*. Englewood Cliffs, N.J.: Prentice-Hall, Inc., 1953.

Meserve, Bruce E., *Fundamental Concepts of Algebra*. Reading, Massachusetts: Addison-Wesley Publishing Co., Inc., 1953.

Meserve, Bruce E., Anthony J. Pettofrezzo, and Dorothy T. Meserve, *Principles of Advanced Mathematics*. Syracuse, New York: The L. W. Singer Company, Inc., 1964.

Murdoch, D. C., *Linear Algebra for Undergraduates*. New York: John Wiley & Sons, Inc., 1957.

Paige, Lowell J., and J. Dean Swift, *Elements of Linear Algebra*. Boston: Ginn & Company, 1961.

Pedoe, Daniel, *A Geometric Introduction to Linear Algebra.* New York: John Wiley & Sons, Inc., 1963.

Perlis, Sam, *Theory of Matrices.* Reading, Massachusetts: Addison-Wesley Publishing Co., Inc., 1952.

Pettofrezzo, Anthony J., *Vectors and Their Applications.* Englewood Cliffs, N.J.: Prentice-Hall, Inc., 1966.

School Mathematics Study Group, *Introduction to Matrix Algebra.* New Haven, Connecticut: Yale University Press, 1961.

Wade, Thomas L., *The Algebra of Vectors and Matrices.* Reading, Massachusetts: Addison-Wesley Publishing Co., Inc., 1951.

Answers to
Odd-Numbered Exercises

1-1 Definitions and Elementary Properties

1.
$$\begin{pmatrix} 0 & 2 & 4 \\ 3 & 5 & 7 \\ 8 & 10 & 12 \end{pmatrix}.$$

3. (a) The elements are in row two; (b) the elements are in column one; (c) the elements are those in the upper left-hand corner and the lower right-hand corner.

5.
$$(A + B) + C = \begin{pmatrix} 4 & -2 & 1 \\ -2 & 6 & 4 \end{pmatrix} + \begin{pmatrix} 2 & 7 & -1 \\ -2 & 1 & 3 \end{pmatrix} = \begin{pmatrix} 6 & 5 & 0 \\ -4 & 7 & 7 \end{pmatrix};$$

$$A + (B + C) = \begin{pmatrix} 3 & 1 & 1 \\ -2 & 5 & 0 \end{pmatrix} + \begin{pmatrix} 3 & 4 & -1 \\ -2 & 2 & 7 \end{pmatrix} = \begin{pmatrix} 6 & 5 & 0 \\ -4 & 7 & 7 \end{pmatrix};$$

$$(A + B) + C = A + (B + C).$$

7.
$$\begin{pmatrix} a_{11} & a_{12} \\ a_{21} & a_{22} \end{pmatrix} = \begin{pmatrix} -6 & 1 \\ 5 & 0 \end{pmatrix}.$$

1-2 Matrix Multiplication

1.
$$AB = \begin{pmatrix} 18 & 1 & 26 \\ -8 & 7 & 6 \end{pmatrix}; \quad BA \text{ does not exist.}$$

3.
$$A(BC) = \begin{pmatrix} 2 & 1 \\ -1 & 0 \end{pmatrix}\begin{pmatrix} 2 & 4 \\ -4 & 0 \end{pmatrix} = \begin{pmatrix} 0 & 8 \\ -2 & -4 \end{pmatrix};$$

$$(AB)C = \begin{pmatrix} 8 & 0 \\ -3 & -1 \end{pmatrix}\begin{pmatrix} 0 & 1 \\ 2 & 1 \end{pmatrix} = \begin{pmatrix} 0 & 8 \\ -2 & -4 \end{pmatrix};$$

$$A(BC) = (AB)C.$$

5.
$$\begin{pmatrix} 1 & 0 \\ 0 & 1 \end{pmatrix}.$$

7.
$$(A + B)(A - B) = \begin{pmatrix} 3 & 4 \\ -3 & 2 \end{pmatrix}\begin{pmatrix} -1 & 2 \\ -1 & -2 \end{pmatrix} = \begin{pmatrix} -7 & -2 \\ 1 & -10 \end{pmatrix};$$

$$A^2 - B^2 = \begin{pmatrix} -5 & 3 \\ -2 & -6 \end{pmatrix} - \begin{pmatrix} 3 & 4 \\ -4 & 3 \end{pmatrix} = \begin{pmatrix} -8 & -1 \\ 2 & -9 \end{pmatrix};$$

$$(A + B)(A - B) \neq A^2 - B^2.$$

9. (a) No matrices exist;
   (b)
$$\begin{pmatrix} r & s \\ 3s & r \end{pmatrix},$$

where r and s are arbitrary scalars.

11.
$$\begin{pmatrix} 20 & 21 \\ 7 & 6 \end{pmatrix}.$$

13.

$$
\begin{aligned}
C(AB + BA) &= C(AB) + C(BA) && \text{distributive property} \\
&= (CA)B + (CB)A && \text{associative property of multiplication} \\
&= (AC)B + (BC)A && AC = CA \text{ and } BC = CB \\
&= A(CB) + B(CA) && \text{associative property of multiplication} \\
&= A(BC) + B(AC) && AC = CA \text{ and } BC = CB \\
&= (AB)C + (BA)C && \text{associative property of multiplication} \\
&= (AB + BA)C && \text{distributive property.}
\end{aligned}
$$

1-3 Diagonal Matrices

1.
$$\begin{pmatrix} 4 & 3 & 3 & 3 \\ 6 & 10 & 6 & 6 \\ 9 & 9 & 18 & 9 \end{pmatrix}.$$

3.
$$\begin{pmatrix} -1 & -1 & -1 \\ 0 & 1 & 0 \\ 0 & 0 & 1 \end{pmatrix}\begin{pmatrix} -1 & -1 & -1 \\ 0 & 1 & 0 \\ 0 & 0 & 1 \end{pmatrix}$$

$$= \begin{pmatrix} 1 + 0 + 0 & 1 - 1 + 0 & 1 + 0 - 1 \\ 0 + 0 + 0 & 0 + 1 + 0 & 0 + 0 + 0 \\ 0 + 0 + 0 & 0 + 0 + 0 & 0 + 0 + 1 \end{pmatrix} = \begin{pmatrix} 1 & 0 & 0 \\ 0 & 1 & 0 \\ 0 & 0 & 1 \end{pmatrix} = I.$$

1-4 Special Real Matrices

1. (a) C, D, and J;   (b) F and J.

3. $n^2 - n + 1$.

5.
$$EE^T = \begin{pmatrix} 5 & 2 & 1 \\ 4 & 2 & 4 \\ 1 & 2 & 3 \end{pmatrix}\begin{pmatrix} 5 & 4 & 1 \\ 2 & 2 & 2 \\ 1 & 4 & 3 \end{pmatrix} = \begin{pmatrix} 30 & 28 & 12 \\ 28 & 36 & 20 \\ 12 & 20 & 14 \end{pmatrix} = (EE^T)^T.$$

7.
$$E - E^T = \begin{pmatrix} 5 & 2 & 1 \\ 4 & 2 & 4 \\ 1 & 2 & 3 \end{pmatrix} - \begin{pmatrix} 5 & 4 & 1 \\ 2 & 2 & 2 \\ 1 & 4 & 3 \end{pmatrix} = \begin{pmatrix} 0 & -2 & 0 \\ 2 & 0 & 2 \\ 0 & -2 & 0 \end{pmatrix};$$

$$(E - E^T)^T = \begin{pmatrix} 0 & 2 & 0 \\ -2 & 0 & -2 \\ 0 & 2 & 0 \end{pmatrix}; \quad (E - E^T)^T = -(E - E^T).$$

9. Let A be any skew-symmetric matrix. Then $A = -A^T$ and $A^2 = (-A^T)(-A^T)$ $= A^T A^T = (AA)^T = (A^2)^T$. Hence, A^2 is a symmetric matrix.

11. Let A and B be skew-symmetric matrices of the same order such that AB is a symmetric matrix. Then $AB = (AB)^T = B^T A^T = (-B)(-A) = BA$. To prove the converse, let A and B be skew-symmetric matrices of the same order such that $AB = BA$. Then $AB = BA = (-B^T)(-A^T) = B^T A^T = (AB)^T$; that is, AB is a symmetric matrix.

1-5 Special Complex Matrices

1. (a) $\begin{pmatrix} 1 & 3 - 2i \\ i & 2 + i \end{pmatrix}$;   (b) $\begin{pmatrix} 1 & i \\ 3 - 2i & 2 + i \end{pmatrix}.$

3. Let $A = ((a_{ij} + b_{ij}i))$ be a skew-Hermitian matrix. Then $A = -A^* = -(\bar{A})^T$. Since the diagonal elements of A and $-(\bar{A})^T$ are equal, $a_{ii} + b_{ii}i$ $= -(a_{ii} - b_{ii}i)$; $a_{ii} = -a_{ii}$; and $a_{ii} = 0$. Therefore, the diagonal elements of A are of the form $b_{ii}i$. Hence, the diagonal elements of a skew-Hermitian matrix are either zeros or pure imaginary numbers.

5. $(A + B)^* = (\overline{A + B})^T = (\bar{A} + \bar{B})^T = (\bar{A})^T + (\bar{B})^T = A^* + B^*.$

7. $(AA^*)^* = (A^*)^* A^* = AA^*.$

9. $(A - A^*)^* = A^* - (A^*)^* = A^* - A; \ A - A^* = -(A - A^*)^*.$

11. Let $((h_{ij}))$ be any Hermitian matrix where $h_{ij} = a_{ij} + b_{ij}i$. Since $((h_{ij})) = ((\overline{h_{ij}}))^T$, then $a_{ij} + b_{ij}i = a_{ji} - b_{ji}i$ for all pairs (i, j); $a_{ij} = a_{ji}$ and $b_{ij} = -b_{ji}$. Hence, $((h_{ij})) = ((a_{ij})) + ((b_{ij}))i$ where $((a_{ij}))$ is a real symmetric matrix and $((b_{ij}))$ is a real skew-symmetric matrix.

CHAPTER 2—INVERSES AND SYSTEMS OF MATRICES

2-1 Determinants

1. 14.   **3.** -2.   **5.** 27.

7. By Theorem 2-5, the elements of one of the two identical rows can be made zero by adding to each element the product of (-1) and the corresponding element of the other row. Then, by Theorem 2-3, the determinant is zero.

9. $k^4 m$.

11.
$$AB = \begin{pmatrix} 7 & 36 \\ 7 & 39 \end{pmatrix};$$

$\det AB = 21 = (-7)(-3) = \det A \det B$.

13.
$$\begin{vmatrix} 5 - \lambda & 1 \\ 2 & 3 - \lambda \end{vmatrix} = \lambda^2 - 8\lambda + 13 = 0;$$

$$\begin{pmatrix} 5 & 1 \\ 2 & 3 \end{pmatrix}\begin{pmatrix} 5 & 1 \\ 2 & 3 \end{pmatrix} - 8\begin{pmatrix} 5 & 1 \\ 2 & 3 \end{pmatrix} + 13\begin{pmatrix} 1 & 0 \\ 0 & 1 \end{pmatrix}$$

$$= \begin{pmatrix} 27 & 8 \\ 16 & 11 \end{pmatrix} + \begin{pmatrix} -40 & -8 \\ -16 & -24 \end{pmatrix} + \begin{pmatrix} 13 & 0 \\ 0 & 13 \end{pmatrix} = \begin{pmatrix} 0 & 0 \\ 0 & 0 \end{pmatrix}.$$

15.
$$\begin{vmatrix} a+b & a & a \\ a & a+b & a \\ a & a & a+b \end{vmatrix} = \begin{vmatrix} b & 0 & -b \\ a & a+b & a \\ a & a & a+b \end{vmatrix} = \begin{vmatrix} b & 0 & 0 \\ a & a+b & 2a \\ a & a & 2a+b \end{vmatrix}$$

$$= b\begin{vmatrix} a+b & 2a \\ a & 2a+b \end{vmatrix} = b(3ab + b^2) = b^2(3a + b).$$

2-2 Inverse of a Matrix

1. $\begin{pmatrix} 2 & -1 \\ -\frac{5}{2} & \frac{3}{2} \end{pmatrix}$.

3. $\begin{pmatrix} \cos\theta & \sin\theta \\ -\sin\theta & \cos\theta \end{pmatrix}$.

5. $\begin{pmatrix} 1 & -3 & 2 \\ -3 & 3 & -1 \\ 2 & -1 & 0 \end{pmatrix}$.

7. $\begin{pmatrix} x \\ y \end{pmatrix} = \begin{pmatrix} \frac{4}{5} & -\frac{1}{5} \\ -\frac{3}{5} & \frac{2}{5} \end{pmatrix}\begin{pmatrix} 4 \\ 1 \end{pmatrix} = \begin{pmatrix} 3 \\ -2 \end{pmatrix};$

that is, $x = 3$ and $y = -2$.

9. A left multiplicative inverse is of the form

$$\begin{pmatrix} 4 & -1 & r \\ -3 & 1 & s \end{pmatrix},$$

where r and s are arbitrary scalars. A right multiplicative inverse does not exist since

$$\begin{pmatrix} 1 & 1 \\ 3 & 4 \\ 0 & 0 \end{pmatrix}\begin{pmatrix} a & b & c \\ d & e & f \end{pmatrix} = \begin{pmatrix} a+d & b+e & c+f \\ 3a+4d & 3b+4e & 3c+4f \\ 0 & 0 & 0 \end{pmatrix} \neq \begin{pmatrix} 1 & 0 & 0 \\ 0 & 1 & 0 \\ 0 & 0 & 1 \end{pmatrix}$$

for any values of a, b, c, d, e, and f.

11. Assume A is a nonsingular matrix. Then A^{-1} exists and $A^{-1}(AB) = A^{-1}0$; $(A^{-1}A)B = 0$; $IB = 0$; $B = 0$, which is contrary to the hypothesis. Hence, A is a singular matrix.

13.
$$AB = \begin{pmatrix} 4 & 4 \\ 3 & 4 \end{pmatrix}; \quad (AB)^{-1} = \begin{pmatrix} 1 & -1 \\ -\frac{3}{4} & 1 \end{pmatrix};$$

$$B^{-1}A^{-1} = \begin{pmatrix} 1 & 0 \\ \frac{1}{2} & \frac{1}{2} \end{pmatrix}\begin{pmatrix} 1 & -1 \\ -\frac{5}{2} & 3 \end{pmatrix} = \begin{pmatrix} 1 & -1 \\ -\frac{3}{4} & 1 \end{pmatrix}; \quad (AB)^{-1} = B^{-1}A^{-1}.$$

15. Let A be any nonsingular symmetric matrix. Then $A^{-1}A = AA^{-1} = I = (AA^{-1})^T = (A^{-1})^T A^T = (A^{-1})^T A$. Hence, $A^{-1} = (A^{-1})^T$; that is, A^{-1} is a symmetric matrix.

17. Let A be any nonsingular matrix. Then $AA^{-1} = I = (AA^{-1})^T = (A^{-1})^T A^T$. Hence, $(A^T)^{-1} = (A^{-1})^T$.

2-3 Systems of Matrices

1. Not a ring.   **3.** A ring.   **5.** A ring.

7.
$$\begin{pmatrix} 5 & -3 \\ 3 & 5 \end{pmatrix} + \begin{pmatrix} 2 & -4 \\ 4 & 2 \end{pmatrix} = \begin{pmatrix} 7 & -7 \\ 7 & 7 \end{pmatrix} \longleftrightarrow 7 - 7i = (5 - 3i) + (2 - 4i).$$

9.
$$\begin{pmatrix} 0 & 1 \\ -1 & 0 \end{pmatrix}\begin{pmatrix} 0 & 1 \\ -1 & 0 \end{pmatrix} = \begin{pmatrix} -1 & 0 \\ 0 & -1 \end{pmatrix} \longleftrightarrow -1 = (i) \times (i).$$

11. $A \odot B = AB - BA = -(BA - AB) = -B \odot A.$

13.
$$\begin{pmatrix} 5i & 1+i \\ -1+i & -5i \end{pmatrix}\begin{pmatrix} 3+i & 2+4i \\ -2+4i & 3-i \end{pmatrix} = \begin{pmatrix} -11+17i & -16+12i \\ 16+12i & -11-17i \end{pmatrix} \longleftrightarrow$$
$$-11 + 17i - 16j + 12k = (5i + j + k) \times (3 + i + 2j + 4k).$$

2-4 Rank of a Matrix

1. $t(2x + z) - 2t(x + y) + t(2y - z) = 0$ for any nonzero scalar t.

3. Three.   **5.** Two.   **7.** $\dfrac{n(n+1)}{2}$.

9.
$$\begin{pmatrix} 1 & 0 \\ 0 & 3 \end{pmatrix}\begin{pmatrix} 1 & -1 \\ 0 & 1 \end{pmatrix}\begin{pmatrix} 1 & 0 \\ -2 & 1 \end{pmatrix}\begin{pmatrix} \frac{1}{6} & 0 \\ 0 & 1 \end{pmatrix} = \begin{pmatrix} \frac{1}{2} & -1 \\ -1 & 3 \end{pmatrix}.$$

Other elementary row transformation matrices exist whose product is

$$\begin{pmatrix} \frac{1}{2} & -1 \\ -1 & 3 \end{pmatrix}.$$

11.

$$\begin{pmatrix} 1 & 0 & 2 \\ 0 & 1 & 0 \\ 0 & 0 & 1 \end{pmatrix}\begin{pmatrix} 1 & 0 & 0 \\ 0 & 1 & 1 \\ 0 & 0 & 1 \end{pmatrix}\begin{pmatrix} 1 & 0 & 0 \\ 0 & 1 & 0 \\ 0 & 0 & \frac{1}{2} \end{pmatrix}\begin{pmatrix} 1 & 0 & 0 \\ 0 & 1 & 0 \\ 0 & 1 & 1 \end{pmatrix}\begin{pmatrix} 1 & 2 & 0 \\ 0 & 1 & 0 \\ 0 & 0 & 1 \end{pmatrix}$$

$$\begin{pmatrix} 1 & 0 & 0 \\ 0 & -1 & 0 \\ 0 & 0 & 1 \end{pmatrix}\begin{pmatrix} 1 & 0 & 0 \\ 0 & 1 & 0 \\ 5 & 0 & 1 \end{pmatrix}\begin{pmatrix} 1 & 0 & 0 \\ 2 & 1 & 0 \\ 0 & 0 & 1 \end{pmatrix} = \begin{pmatrix} 0 & -3 & 1 \\ -\frac{1}{2} & -\frac{3}{2} & \frac{1}{2} \\ \frac{3}{2} & -\frac{1}{2} & \frac{1}{2} \end{pmatrix}.$$

Other elementary row transformation matrices exist whose product is

$$\begin{pmatrix} 0 & -3 & 1 \\ -\frac{1}{2} & -\frac{3}{2} & \frac{1}{2} \\ \frac{3}{2} & -\frac{1}{2} & \frac{1}{2} \end{pmatrix}.$$

2-5 Systems of Linear Equations

1. An infinite number of solutions of the form $x = -\frac{15}{14}z - \frac{17}{7}$ and $y = \frac{5}{7}z - \frac{26}{7}$ exists.

3. $k = 18$.

5. A nontrivial solution does not exist.

CHAPTER 3—TRANSFORMATIONS OF THE PLANE

3-1 Mappings

1. (a) T is a one-to-one mapping of R onto R; (b) T is not a mapping of R onto R; (c) T is a one-to-one mapping of R onto R; (d) T is a mapping of R onto R.

3. $n!$

5. A single-valued mapping T^{-1} does not exist such that $T^{-1}(b) = 1$ and $T^{-1}(b) = 3$.

3-2 Rotations

1. $(3, 2)$.   **3.** $(1, \sqrt{3})$.   **5.** $\left(-\dfrac{\sqrt{2}}{2}, \dfrac{3\sqrt{2}}{2}\right).$

7. $3x^2 + y^2 = 32$.   **9.** $x^2 + y^2 = r^2$.

11. Let

$$\begin{pmatrix} \cos\theta & -\sin\theta \\ \sin\theta & \cos\theta \end{pmatrix} \quad \text{and} \quad \begin{pmatrix} \cos\phi & -\sin\phi \\ \sin\phi & \cos\phi \end{pmatrix}$$

be any two rotation matrices of the form (3-3).

(a) $\begin{pmatrix} \cos\theta & -\sin\theta \\ \sin\theta & \cos\theta \end{pmatrix}\begin{pmatrix} \cos\phi & -\sin\phi \\ \sin\phi & \cos\phi \end{pmatrix} = \begin{pmatrix} \cos(\theta+\phi) & -\sin(\theta+\phi) \\ \sin(\theta+\phi) & \cos(\theta+\phi) \end{pmatrix}$,

a rotation matrix of the form (3-3);

(b) $\begin{pmatrix} \cos(\theta + \phi) & -\sin(\theta + \phi) \\ \sin(\theta + \phi) & \cos(\theta + \phi) \end{pmatrix} = \begin{pmatrix} \cos(\phi + \theta) & -\sin(\phi + \theta) \\ \sin(\phi + \theta) & \cos(\phi + \theta) \end{pmatrix}$

$$= \begin{pmatrix} \cos\phi & -\sin\phi \\ \sin\phi & \cos\phi \end{pmatrix} \begin{pmatrix} \cos\theta & -\sin\theta \\ \sin\theta & \cos\theta \end{pmatrix}.$$

13. Let (x_1, y_1) and (x_2, y_2) be any two points on a coordinate plane. The distance d between these two points is equal to $\sqrt{(x_2 - x_1)^2 + (y_2 - y_1)^2}$. Under any rotation of the plane about the origin represented by

$$\begin{pmatrix} \cos\theta & -\sin\theta \\ \sin\theta & \cos\theta \end{pmatrix},$$

these points are mapped onto $(x_1 \cos\theta - y_1 \sin\theta, x_1 \sin\theta + y_1 \cos\theta)$ and $(x_2 \cos\theta - y_2 \sin\theta, x_2 \sin\theta + y_2 \cos\theta)$, respectively. The distance d' between these image points is equal to

$$\{[(x_2 \cos\theta - y_2 \sin\theta) - (x_1 \cos\theta - y_1 \sin\theta)]^2$$
$$+ [(x_2 \sin\theta + y_2 \cos\theta) - (x_1 \sin\theta + y_1 \cos\theta)]^2\}^{1/2}$$
$$= \sqrt{[(x_2 - x_1)\cos\theta - (y_2 - y_1)\sin\theta]^2 + [(x_2 - x_1)\sin\theta + (y_2 - y_1)\cos\theta]^2}$$
$$= \sqrt{(x_2 - x_1)^2(\cos^2\theta + \sin^2\theta) + (y_2 - y_1)^2(\sin^2\theta + \cos^2\theta)}$$
$$= \sqrt{(x_2 - x_1)^2 + (y_2 - y_1)^2}.$$

Hence, $d = d'$ and the distance between two points on a plane is invariant under a rotation of the plane about the origin.

3-3 Reflections, Dilations, and Magnifications

1. Each point (x, y) is mapped onto the point $(2x, 2y)$; that is, the matrix represents a dilation of the plane. The transformation is an example of a one-to-one mapping of the set of points on the plane onto itself.

3. Each point (x, y) is mapped onto the point $(2y, 2x)$; that is, the matrix represents the product of a dilation of the plane and a reflection of the plane with respect to the line $y = x$. The transformation is an example of a one-to-one mapping of the set of points on the plane onto itself.

5. Let (x_1, y_1) and (x_2, y_2) be any two points on a coordinate plane. The distance d between these two points is equal to $\sqrt{(x_2 - x_1)^2 + (y_2 - y_1)^2}$.

   (a) Under a reflection of the plane with respect to the x-axis represented by

$$\begin{pmatrix} 1 & 0 \\ 0 & -1 \end{pmatrix},$$

these points are mapped onto $(x_1, -y_1)$ and $(x_2, -y_2)$, respectively. The distance d' between these image points is equal to

$$\sqrt{(x_2 - x_1)^2 + (-y_2 + y_1)^2} = \sqrt{(x_2 - x_1)^2 + (y_2 - y_1)^2}.$$

Hence, $d = d'$ and the distance between two points on a plane is invariant under a reflection of the plane with respect to the x-axis.

   (b) Under a reflection of the plane with respect to the y-axis represented by

$$\begin{pmatrix} -1 & 0 \\ 0 & 1 \end{pmatrix},$$

these points are mapped onto $(-x_1, y_1)$ and $(-x_2, y_2)$, respectively. The distance d'' between these image points is equal to

$$\sqrt{(-x_2 + x_1)^2 + (y_2 - y_1)^2} = \sqrt{(x_2 - x_1)^2 + (y_2 - y_1)^2}.$$

Hence, $d = d''$ and the distance between two points on a plane is invariant under a reflection of the plane with respect to the y-axis.

7. The multiplication of any matrix A and a conformable scalar matrix is commutative. Hence, the multiplication of any dilation matrix (a scalar matrix) and any rotation matrix of the form (3-3) is commutative.

9. **(a)** $\begin{pmatrix} 4 & 0 \\ 0 & 4 \end{pmatrix};$ **(b)** $\begin{pmatrix} \frac{1}{2} & 0 \\ 0 & \frac{1}{2} \end{pmatrix};$ **(c)** $\begin{pmatrix} r & 0 \\ 0 & r \end{pmatrix}.$

11. The line $2x + 5y = 10$ is mapped onto the line $x + y = 1$.

3-4 Other Transformations

1. Each point (x, y) is mapped onto a point $(0, y)$; that is, the matrix represents a vertical projection of the points on the plane onto the y-axis. The transformation is a mapping of the set of points on the plane into itself.

3. Each point (x, y) is mapped onto a point $(x - 2y, y)$; that is, the matrix represents a shear parallel to the x-axis. The transformation is a one-to-one mapping of the set of points on the plane onto itself.

5. **(a)** The circle is mapped onto the ellipse $5x^2 - 4xy + y^2 = 1$.
   (b) The rectangle is mapped onto a parallelogram with vertices at $(0, 0)$, $(2, 2k)$, $(2, 2k + 1)$, and $(0, 1)$.

7.

$$\begin{pmatrix} 2 & 3 \\ 1 & 2 \end{pmatrix}.$$

3-5 Linear Homogeneous Transformations

In Exercises 1 and 3 let

$$T = \begin{pmatrix} a & b \\ c & d \end{pmatrix}$$

be a nonsingular matrix.

1.

$$\begin{pmatrix} a & b \\ c & d \end{pmatrix}\begin{pmatrix} 0 \\ 0 \end{pmatrix} = \begin{pmatrix} 0 \\ 0 \end{pmatrix}.$$

3. Let $Ax + By + C = 0$ and $Ax + By + D = 0$ be the equations of parallel lines. By the results of Exercise 2, the images of the parallel lines under the transformation represented by T are the lines

$$\frac{Ad - Bc}{ad - bc}x + \frac{Ba - Ab}{ad - bc}y + C = 0$$

and

$$\frac{Ad - Bc}{ad - bc}x + \frac{Ba - Ab}{ad - bc}y + D = 0,$$

a pair of parallel lines. Hence, the images of parallel lines are parallel lines.

5. Under the *singular* homogeneous transformation represented by

$$\begin{pmatrix} 6 & 3 \\ 2 & 1 \end{pmatrix},$$

the image (x', y') of each point (x, y) is such that $x' = 6x + 3y$ and $y' = 2x + y$; that is, $x' = 3y'$, or $x = 3y$. Hence, the set of points on the plane are mapped onto the line $x = 3y$.

3-6 Orthogonal Matrices

1. A proper orthogonal matrix.

3. Neither a proper orthogonal matrix nor an improper orthogonal matrix.

5.

$$(A^{-1})(A^{-1})^T = \begin{pmatrix} \frac{2}{3} & \frac{1}{3} & \frac{2}{3} \\ -\frac{2}{3} & \frac{2}{3} & \frac{1}{3} \\ \frac{1}{3} & \frac{2}{3} & -\frac{2}{3} \end{pmatrix}\begin{pmatrix} \frac{2}{3} & -\frac{2}{3} & \frac{1}{3} \\ \frac{1}{3} & \frac{2}{3} & \frac{2}{3} \\ \frac{2}{3} & \frac{1}{3} & -\frac{2}{3} \end{pmatrix} = I.$$

7.

$$\det\left[\begin{pmatrix} \frac{12}{13} & \frac{5}{13} \\ \frac{5}{13} & -\frac{12}{13} \end{pmatrix} + \begin{pmatrix} 1 & 0 \\ 0 & 1 \end{pmatrix}\right] = \det\begin{pmatrix} \frac{25}{13} & \frac{5}{13} \\ \frac{5}{13} & \frac{1}{13} \end{pmatrix} = \frac{25}{13} \cdot \frac{1}{13} - \frac{5}{13} \cdot \frac{5}{13} = 0.$$

9.

$$\begin{pmatrix} 1 & 0 \\ 0 & 1 \end{pmatrix} \quad \text{and} \quad \begin{pmatrix} 0 & 1 \\ 1 & 0 \end{pmatrix}.$$

3-7 Translations

1. In **(a)** through **(d)**, k is any nonzero real number.
(a) $(k, -2k, k)$;   **(b)** $(3k, 0, k)$;
(c) $(0, 0, k)$;   **(d)** $(3k, 4k, k)$.

3. (a)
$$\begin{pmatrix} x' \\ y' \\ 1 \end{pmatrix} = \begin{pmatrix} 1 & 0 & -2 \\ 0 & 1 & 4 \\ 0 & 0 & 1 \end{pmatrix}\begin{pmatrix} x \\ y \\ 1 \end{pmatrix};$$
(b)
$$\begin{pmatrix} x' \\ y' \\ 1 \end{pmatrix} = \begin{pmatrix} 1 & 0 & 0 \\ 0 & 1 & 1 \\ 0 & 0 & 1 \end{pmatrix}\begin{pmatrix} x \\ y \\ 1 \end{pmatrix}.$$

5. $(8, -2)$.   **7.** $(10, 4)$.   **9.** $3x^2 + 2y^2 - 18 = 0$.

11. Let (x_1, y_1) and (x_2, y_2) be any two points on a plane. The distance d between these two points is equal to $\sqrt{(x_2 - x_1)^2 + (y_2 - y_1)^2}$. Under any translation of the plane represented by

$$\begin{pmatrix} 1 & 0 & a \\ 0 & 1 & b \\ 0 & 0 & 1 \end{pmatrix},$$

the points are mapped onto $(x_1 + a, y_1 + b)$ and $(x_2 + a, y_2 + b)$, respectively. The distance d' between these image points is equal to

$\sqrt{[(x_2 + a) - (x_1 + a)]^2 + [(y_2 + b) - (y_1 + b)]^2}$; that is,
$\sqrt{(x_2 - x_1)^2 + (y_2 - y_1)^2}$. Hence, $d = d'$ and the distance between two points on a plane is invariant under a translation of the plane.

13.
$$\left(\frac{\sqrt{3} - 6}{2}, \frac{1}{2}\right).$$

3-8 Rigid Motion Transformations

1.
$$\mathcal{R} = \begin{pmatrix} \frac{1}{2} & -\frac{\sqrt{3}}{2} & 2 \\ \frac{\sqrt{3}}{2} & \frac{1}{2} & 0 \\ 0 & 0 & 1 \end{pmatrix}; \quad (2, 0, 1).$$

3.
$$a_{31} = a_{32} = 0, \quad a_{33} = 1, \quad \begin{vmatrix} a_{11} & a_{12} \\ a_{21} & a_{22} \end{vmatrix} = \begin{vmatrix} \frac{1}{2} & -\frac{\sqrt{3}}{2} \\ \frac{\sqrt{3}}{2} & \frac{1}{2} \end{vmatrix} = 1;$$

$$\left(\frac{1}{2}\right)^2 + \left(\frac{\sqrt{3}}{2}\right)^2 = 1, \quad \left(-\frac{\sqrt{3}}{2}\right)^2 + \left(\frac{1}{2}\right)^2 = 1;$$

$$\left(\frac{1}{2}\right)\left(-\frac{\sqrt{3}}{2}\right) + \left(\frac{\sqrt{3}}{2}\right)\left(\frac{1}{2}\right) = 0.$$

5. (a) $\begin{pmatrix} -1 & 0 & 6 \\ 0 & 1 & 0 \\ 0 & 0 & 1 \end{pmatrix}$;
   (b) $\begin{pmatrix} \frac{1}{2} & \frac{\sqrt{3}}{2} & -\frac{1}{2} \\ \frac{\sqrt{3}}{2} & -\frac{1}{2} & \frac{\sqrt{3}}{2} \\ 0 & 0 & 0 \end{pmatrix}.$

7. The matrices representing a rotation of the plane about the origin are of the form
$$R = \begin{pmatrix} \cos\theta & -\sin\theta & 0 \\ \sin\theta & \cos\theta & 0 \\ 0 & 0 & 1 \end{pmatrix}.$$

The reflection matrices of (3-6) are of the form
$$F = \begin{pmatrix} \mp 1 & 0 & 0 \\ 0 & \pm 1 & 0 \\ 0 & 0 & 1 \end{pmatrix}. \quad \text{Then} \quad RF = \begin{pmatrix} \mp\cos\theta & \mp\sin\theta & 0 \\ \mp\sin\theta & \pm\cos\theta & 0 \\ 0 & 0 & 1 \end{pmatrix},$$

where $a_{31} = a_{32} = 0$, $a_{33} = 1$,
$$\begin{vmatrix} a_{11} & a_{12} \\ a_{21} & a_{22} \end{vmatrix} = \begin{vmatrix} \mp\cos\theta & \mp\sin\theta \\ \mp\sin\theta & \pm\cos\theta \end{vmatrix} = -1;$$

$(\mp\cos\theta)^2 + (\mp\sin\theta)^2 = 1$, $(\mp\sin\theta)^2 + (\pm\cos\theta)^2 = 1$; and $(\mp\cos\theta)(\mp\sin\theta)$ $+ (\mp\sin\theta)(\pm\cos\theta) = 0$. Similarly,

$$FR = \begin{pmatrix} \mp\cos\theta & \pm\sin\theta & 0 \\ \pm\sin\theta & \pm\cos\theta & 0 \\ 0 & 0 & 1 \end{pmatrix},$$

where $a_{31} = a_{32} = 0$, $a_{33} = 1$,

$$\begin{vmatrix} a_{11} & a_{12} \\ a_{21} & a_{22} \end{vmatrix} = \begin{vmatrix} \mp\cos\theta & \pm\sin\theta \\ \pm\sin\theta & \pm\cos\theta \end{vmatrix} = -1;$$

$(\mp\cos\theta)^2 + (\pm\sin\theta)^2 = 1$, $(\pm\sin\theta)^2 + (\pm\cos\theta)^2 = 1$; and $(\mp\cos\theta)(\pm\sin\theta) + (\pm\sin\theta)(\pm\cos\theta) = 0$.

Chapter 4—Eigenvalues and Eigenvectors

4-1 Characteristic Functions

1. $\lambda^2 - 9\lambda + 14 = 0$; $\lambda_1 = 2$ and $\lambda_2 = 7$; $(k \quad -k)^T$ and $(3k \quad 2k)^T$, where k is any nonzero scalar.

3. $\lambda^2 - 2\lambda = 0$; $\lambda_1 = 0$ and $\lambda_2 = 2$; $(0 \quad k)^T$ and $(k \quad 0)^T$, where k is any nonzero scalar.

5. $\lambda^3 - 2\lambda^2 - 5\lambda + 6 = 0$; $\lambda_1 = 1$, $\lambda_2 = -2$, and $\lambda_3 = 3$; $(k \quad -k \quad -k)^T$, $(11k \quad k \quad -14k)^T$, and $(k \quad k \quad k)^T$, where k is any nonzero scalar.

7. (a) Let

$$A = \begin{pmatrix} 5 & 3 \\ 2 & 4 \end{pmatrix}. \quad \text{Then} \quad A^2 = \begin{pmatrix} 31 & 27 \\ 18 & 22 \end{pmatrix},$$

$t_1 = 9$, and $t_2 = 53$. By (4-4), $c_0 = 1$, $c_1 = -9$, and $c_2 = 14$. Hence, $\lambda^2 - 9\lambda + 14 = 0$.

(b) Let

$$A = \begin{pmatrix} 2 & -2 & 3 \\ 1 & 1 & 1 \\ 1 & 3 & -1 \end{pmatrix}. \quad \text{Then} \quad A^2 = \begin{pmatrix} 5 & 3 & 1 \\ 4 & 2 & 3 \\ 4 & -2 & 7 \end{pmatrix},$$

$$A^3 = \begin{pmatrix} 14 & -4 & 17 \\ 13 & 3 & 11 \\ 13 & 11 & 3 \end{pmatrix},$$

$t_1 = 2$, $t_2 = 14$, and $t_3 = 20$. By (4-4), $c_0 = 1$, $c_1 = -2$, $c_2 = -5$, and $c_3 = 6$. Hence, $\lambda^3 - 2\lambda^2 - 5\lambda + 6 = 0$.

9. Every eigenvalue of A satisfies the characteristic equation $f(\lambda) = |A - \lambda I| = 0$. If $\lambda = 0$, then $|A - 0I| = |A| = 0$; that is, the value of the determinant of A is zero.

11. (a) $k\lambda_1$, $k\lambda_2$, and $k\lambda_3$.   (b) $\lambda_1 - k$, $\lambda_2 - k$, and $\lambda_3 - k$.

13. Let $A = ((\delta_{ij}a_{ij}))$ be any diagonal matrix of order n. Then the characteristic equation of A is $|A - \lambda I| = (a_{11} - \lambda)(a_{22} - \lambda) \cdots (a_{nn} - \lambda) = 0$. Therefore, $\lambda_1 = a_{11}$, $\lambda_2 = a_{22}, \ldots, \lambda_n = a_{nn}$; that is, the eigenvalues of a diagonal matrix are equal to the diagonal elements.

4-2 A Geometric Interpretation of Eigenvectors

1. The one-dimensional vector spaces containing the sets of vectors of the forms $(k \quad 0)^T$ and $(0 \quad k)^T$.

3. The one-dimensional vector space containing the set of vectors of the form $(0 \quad k)^T$.

5. The one-dimensional vector space containing the set of vectors of the form $(k \quad 0)^T$.

7. The characteristic equation of the matrix

$$\begin{pmatrix} \cos \theta & -\sin \theta \\ \sin \theta & \cos \theta \end{pmatrix}$$

representing a rotation of the plane about the origin through an angle θ is $\lambda^2 - 2 \cos \theta \, \lambda + 1 = 0$. Real eigenvalues (and hence real eigenvectors) of A exist if and only if the discriminant of the characteristic equation is greater than or equal to zero; that is, if and only if

$$4 \cos^2 \theta - 4 \geq 0$$
$$\cos^2 \theta - 1 \geq 0$$
$$-\sin^2 \theta \geq 0$$
$$\theta = 180° k,$$

where k is any integer. Therefore, invariant vector spaces exist only under rotations of the plane about the origin through angles which are integral multiples of $180°$.

4-3 Some Theorems

1. Associated with the distinct eigenvalues $\lambda_1 = -1$ and $\lambda_2 = 8$ are the eigenvectors $(5a \quad -4a)^T$ and $(b \quad b)^T$, respectively, where a and b are any nonzero scalars. Then $m_1(5a \quad -4a)^T + m_2(b \quad b)^T = 0$ if and only if $m_1 = m_2 = 0$. Hence, $(5a \quad -4a)^T$ and $(b \quad b)^T$ are linearly independent.

3. Associated with the real, distinct eigenvalues $\lambda_1 = 0$, $\lambda_2 = 1$, and $\lambda_3 = 4$ are the eigenvectors $(a \quad -a \quad 0)^T$, $(0 \quad 0 \quad b)^T$, and $(c \quad c \quad 0)^T$, respectively, where a, b, and c are any nonzero scalars. The eigenvectors are mutually orthogonal since $(a \quad -a \quad 0)(0 \quad 0 \quad b)^T = 0$, $(0 \quad 0 \quad b)(c \quad c \quad 0)^T = 0$, and $(c \quad c \quad 0)(a \quad -a \quad 0)^T = 0$.

5. Let X_i be a unit eigenvector associated with the eigenvalue λ_i of A. Then

$$X_i^T A X_i = (\lambda_i)$$
$$(X_i^T A X_i)^* = (\lambda_i)^*$$
$$(\bar{X}_i)^T A^* \bar{X}_i = (\bar{\lambda}_i);$$

that is, $\bar{\lambda}_i$ is an eigenvalue of A^*. Hence, the eigenvalues of A^* are the conjugates of the eigenvalues of A.

4-4 Diagonalization of Matrices

1. The matrices are not similar matrices.

3. Matrices A and B are similar matrices since any nonsingular matrix of the form

$$\begin{pmatrix} 3c - 2d & 3c - 2d \\ c & d \end{pmatrix},$$

where c and d are arbitrary scalars, is such that

$$\begin{pmatrix} 3c - 2d & 3c - 2d \\ c & d \end{pmatrix}^{-1} \begin{pmatrix} 2 & 0 \\ 1 & 1 \end{pmatrix} \begin{pmatrix} 3c - 2d & 3c - 2d \\ c & d \end{pmatrix} = \begin{pmatrix} 4 & 3 \\ -2 & -1 \end{pmatrix};$$

$\det A = 2 = \det B$; and the eigenvalues of both A and B are $\lambda_1 = 1$ and $\lambda_2 = 2$.

5. Associated with the eigenvalues $\lambda_1 = -1$ and $\lambda_2 = 2$ are the eigenvectors $(k \quad -k)^T$ and $(k \quad 0)^T$, respectively, where k is any nonzero scalar. Since

$$A^3 = \begin{pmatrix} 8 & 9 \\ 0 & -1 \end{pmatrix},$$

$\lambda_1^3 = -1$, and $\lambda_2^3 = 8$, then

$$A^3(k \quad -k)^T = \begin{pmatrix} -k \\ k \end{pmatrix} = \lambda_1^3(k \quad -k)^T$$

and

$$A^3(k \quad 0)^T = \begin{pmatrix} 8k \\ 0 \end{pmatrix} = \lambda_2^3(k \quad 0)^T.$$

7. Any matrix of the form

$$\begin{pmatrix} a & 2b \\ 2a & -3b \end{pmatrix},$$

where a and b are any scalars such that $ab \neq 0$;

$$\begin{pmatrix} \dfrac{3}{7a} & \dfrac{2}{7a} \\ \dfrac{2}{7b} & -\dfrac{1}{7b} \end{pmatrix} \begin{pmatrix} 5 & 4 \\ 12 & 7 \end{pmatrix} \begin{pmatrix} a & 2b \\ 2a & -3b \end{pmatrix} = \begin{pmatrix} 13 & 0 \\ 0 & -1 \end{pmatrix}.$$

9.

$$\begin{pmatrix} \dfrac{1}{\sqrt{2}} & \dfrac{1}{\sqrt{2}} & 0 \\ 0 & 0 & 1 \\ \dfrac{1}{\sqrt{2}} & -\dfrac{1}{\sqrt{2}} & 0 \end{pmatrix} \begin{pmatrix} 2 & 2 & 0 \\ 2 & 2 & 0 \\ 0 & 0 & 1 \end{pmatrix} \begin{pmatrix} \dfrac{1}{\sqrt{2}} & 0 & \dfrac{1}{\sqrt{2}} \\ \dfrac{1}{\sqrt{2}} & 0 & -\dfrac{1}{\sqrt{2}} \\ 0 & 1 & 0 \end{pmatrix} = \begin{pmatrix} 4 & 0 & 0 \\ 0 & 1 & 0 \\ 0 & 0 & 0 \end{pmatrix}.$$

Other orthogonal transformations exist.

11. If A is similar to kI, a nonsingular matrix C exists such that $A = C^{-1}(kI)C$. Then $A = C^{-1}k(IC) = C^{-1}kC = kC^{-1}C = kI$.

4-5 The Hamilton-Cayley Theorem

1. $f(\lambda) = \lambda^2 - 6\lambda + 11$; $f(A) = \begin{pmatrix} 2 & -1 \\ 3 & 4 \end{pmatrix}^2 - 6\begin{pmatrix} 2 & -1 \\ 3 & 4 \end{pmatrix} + 11\begin{pmatrix} 1 & 0 \\ 0 & 1 \end{pmatrix}$

$= \begin{pmatrix} 1 & -6 \\ 18 & 13 \end{pmatrix} + \begin{pmatrix} -12 & 6 \\ -18 & -24 \end{pmatrix} + \begin{pmatrix} 11 & 0 \\ 0 & 11 \end{pmatrix} = \begin{pmatrix} 0 & 0 \\ 0 & 0 \end{pmatrix}.$

3. $f(\lambda) = \lambda^2 - 7\lambda + 10$; $f(A) = \begin{pmatrix} 5 & 0 \\ 0 & 2 \end{pmatrix}^2 - 7\begin{pmatrix} 5 & 0 \\ 0 & 2 \end{pmatrix} + 10\begin{pmatrix} 1 & 0 \\ 0 & 1 \end{pmatrix}$

$= \begin{pmatrix} 25 & 0 \\ 0 & 4 \end{pmatrix} + \begin{pmatrix} -35 & 0 \\ 0 & -14 \end{pmatrix} + \begin{pmatrix} 10 & 0 \\ 0 & 10 \end{pmatrix} = \begin{pmatrix} 0 & 0 \\ 0 & 0 \end{pmatrix}.$

5. $f(\lambda) = \lambda^2 - 6\lambda + 1$; $f(A) = A^2 - 6A + I = 0$, $I = -A^2 + 6A$, and

$$A^{-1} = -A + 6I = -\begin{pmatrix} 5 & 2 \\ 2 & 1 \end{pmatrix} + \begin{pmatrix} 6 & 0 \\ 0 & 6 \end{pmatrix} = \begin{pmatrix} 1 & -2 \\ -2 & 5 \end{pmatrix}.$$

7. (a) $f(\lambda) = \lambda^2 - 8\lambda - 3$; $f(A) = A^2 - 8A - 3I = 0$, $I = \frac{1}{3}(A^2 - 8A)$,

$A^{-1} = \frac{1}{3}(A - 8I)$,

$$A^{-2} = A^{-1}A^{-1} = \frac{1}{3}(I - 8A^{-1}) = \frac{67}{9}I - \frac{8}{9}A$$

$$= \frac{67}{9}\begin{pmatrix} 1 & 0 \\ 0 & 1 \end{pmatrix} - \frac{8}{9}\begin{pmatrix} 7 & 2 \\ 5 & 1 \end{pmatrix} = \begin{pmatrix} \frac{11}{9} & -\frac{16}{9} \\ -\frac{40}{9} & \frac{59}{9} \end{pmatrix};$$

$$A^2A^{-2} = \begin{pmatrix} 59 & 16 \\ 40 & 11 \end{pmatrix}\begin{pmatrix} \frac{11}{9} & -\frac{16}{9} \\ -\frac{40}{9} & \frac{59}{9} \end{pmatrix} = \begin{pmatrix} 1 & 0 \\ 0 & 1 \end{pmatrix}.$$

(b) $A^{-3} = A^{-1}A^{-2} = \frac{67}{9}A^{-1} - \frac{8}{9}I = \frac{67}{27}A - \frac{560}{27}I.$

$$= \frac{67}{27}\begin{pmatrix} 7 & 2 \\ 5 & 1 \end{pmatrix} - \frac{560}{27}\begin{pmatrix} 1 & 0 \\ 0 & 1 \end{pmatrix} = \begin{pmatrix} -\frac{91}{27} & \frac{134}{27} \\ \frac{335}{27} & -\frac{493}{27} \end{pmatrix};$$

$$A^3A^{-3} = \begin{pmatrix} 493 & 134 \\ 335 & 91 \end{pmatrix}\begin{pmatrix} -\frac{91}{27} & \frac{134}{27} \\ \frac{335}{27} & -\frac{493}{27} \end{pmatrix} = \begin{pmatrix} 1 & 0 \\ 0 & 1 \end{pmatrix}.$$

4-6   Quadratic Forms

1. $(x\ \ y)\begin{pmatrix} 3 & 5 \\ 5 & 3 \end{pmatrix}\begin{pmatrix} x \\ y \end{pmatrix}.$ **3.** $(x\ \ y)\begin{pmatrix} 2 & \sqrt{2} \\ \sqrt{2} & 1 \end{pmatrix}\begin{pmatrix} x \\ y \end{pmatrix}.$

5.

$$(x'\ \ y')\begin{pmatrix} \frac{1}{\sqrt{2}} & -\frac{1}{\sqrt{2}} \\ \frac{1}{\sqrt{2}} & \frac{1}{\sqrt{2}} \end{pmatrix}\begin{pmatrix} 1 & -1 \\ -1 & 1 \end{pmatrix}\begin{pmatrix} \frac{1}{\sqrt{2}} & \frac{1}{\sqrt{2}} \\ -\frac{1}{\sqrt{2}} & \frac{1}{\sqrt{2}} \end{pmatrix}\begin{pmatrix} x' \\ y' \end{pmatrix}$$

$$= (x'\ \ y')\begin{pmatrix} 2 & 0 \\ 0 & 0 \end{pmatrix}\begin{pmatrix} x' \\ y' \end{pmatrix} = (2x'^2).$$

Other orthogonal transformations exist.

4-7 Classification of the Conics

1. $3x''^2 + 9y''^2 = 8$, a real ellipse.   **3.** $x''^2 - y''^2 = 4$, a hyperbola.

5. A real ellipse.   **7.** Two intersecting lines.

9. $\lambda_1 = 8$ and $\lambda_2 = 2$; the squares of the semi-axes are $a^2 = 1$ and $b^2 = 4$; $\lambda_1 a^2 = \lambda_2 b^2$.

4-8 Invariants for Conics

1. Three.   **3.** Three.

5. (a) $a + c = 1 + 1 = \frac{1}{2} + \frac{3}{2}$;   **(b)** $ac - b^2 = (1)(1) - (\frac{1}{2})^2 = (\frac{1}{2})(\frac{3}{2})$.

7. (a) $a + c = 7 + (-7) = 1 + (-1)$;
    (b) $ac - b^2 = (7)(-7) - (-24)^2 = (25)(-25)$.

9. (a) The sum of the roots of equation (4-39) is equal to the negative of the coefficient of λ; that is, $a + c = \lambda_1 + \lambda_2$;
   (b) the product of the roots of equation (4-39) is equal to the constant term; that is, $ac - b^2 = \lambda_1 \lambda_2$;
   (c)
$$|F| = \begin{vmatrix} a & b \\ b & c \end{vmatrix} = ac - b^2.$$

11. By the results of this section, $a + c$ is an invariant for conics under a rotation transformation. By (4-28), f is an invariant for conics under a rotation transformation. Hence, $a + c + f$ is an invariant for conics under the rotation transformation.

Index

Addition:
  of matrices, 4
    associative property for, 4
    commutative property for, 4
  of quaternions, 39
Additive identity element:
  for matrices, 4
  for a ring, 35
Additive inverses:
  for matrices, 5
  for a ring, 36
Anti-symmetric matrix, 15
Array, rectangular, 1, 2
Associative property:
  for addition of matrices, 4
  for multiplication of matrices, 10
  for a ring, 35
Augmented matrix, 46

Central conic, 105
Characteristic equation, 83
Characteristic function, 83
Characteristic values, 86; *see also* Eigen-
    values
Characteristic vectors, 86; *see also* Eigen-
    vectors
Classical canonical form, 93

Closure property for a ring, 35
Cofactor, 26
Column:
  index, 2
  matrix, 8
  vector, 8
Commutative property:
  for addition of matrices, 4
  for a ring, 35
Complex matrix, 19
Complex numbers as matrices, 37
Conformable matrices, 8
Conic, 103
  central, 105
  noncentral, 105
  rank of a, 109
Conics:
  degenerate, 110
  invariants for, 109–111
  proper, 110
Conic section, 103
  matrix of a, 104
Conjugate of a matrix, 19
Consistent system of linear equations, 47,
    50
Coordinates:
  homogeneous, 72
  nonhomogeneous, 72

Degenerate conics, 110
Dependent, linearly, 41, 90
Determinant of a matrix:
  of order n, 23–24
  of order three, 22–23
  of order two, 22
  value of a, 22–27
Diagonal, main, 13
Diagonal, principal, 13
Diagonal elements, 13
Diagonalization of a matrix, 93
Diagonal matrix, 13, 61
Difference of two matrices, 5, 16, 17, 21
Dilation of the plane, 61
Distributive property:
  left-hand, 10
  for matrices, 9
  for quaternions, 39
  right-hand, 10
  for a ring, 36

Echelon form, 43, 47
Eigenvalues, 85, 90
  of Hermitian matrices, 90
  of symmetric matrices, 91
Eigenvector, 86
Eigenvectors, 90
  of Hermitian matrices, 91
  of symmetric matrices, 92
Elementary row transformation matrices, 44
Elementary row transformations, 43
Element of a matrix, 2
  cofactor of an, 26
  minor of an, 26
Equal matrices, 3
Equal quaternions, 39

Fibonacci sequence, 12
Fixed point, 56

General linear transformations of the plane, 71, 80, 81

Hamilton-Cayley Theorem, 97
Hermitian matrix, 19
  eigenvalues of a, 90
  eigenvectors of a, 91

Homogeneous:
  coordinates, 72
  linear equations, 49–50
  linear transformations of the plane, 66, 67, 80
  nonsingular, 66, 67
  singular, 122

Identity:
  matrix, 14
  transformation, 56
Image, 52
Improper orthogonal matrix, 69
Inconsistent system of linear equations, 47
Independent, linearly, 41, 90
Index:
  column, 2
  row, 2
Invariant:
  under a transformation, 58, 69–70, 80, 81, 109–110
  vector spaces, 88
Invariants for conics, 109–111
Inverse:
  additive, 5, 36
  mapping, 52
  multiplicative, 29, 32, 45, 99
  left, 28–29
  right, 29
Isomorphic rings, 36
Isomorphism, 36

Kronecker delta, 14

Latent values, 86; see also Eigenvalues
Latent vectors, 86; see also Eigenvectors
Left-hand distributive property, 10
Left multiplicative inverse, 28–29
Linear equations, systems of, 6, 46
  consistent, 47, 50
  homogeneous, 49–50
  inconsistent, 47
Linearly dependent, 41, 90
Linearly independent, 41, 90
Linear transformations of the plane; see also Transformations of the plane:
  general, 71, 80, 81
  homogeneous, 66, 67, 80

Linear transformations (*cont.*):
  homogeneous (*cont.*):
    nonsingular, 66, 67
    singular, 122
    rigid motion, 79–80

Magnification of the plane, 61
Main diagonal, 13
Mapping, single-valued, 51–52, 66
  into, 51–52
  inverse, 52
  one-to-one, 52
  onto, 52
Matrices, 2
  addition of, 4
    associative property for, 4
    commutative property for, 4
  additive identity element for, 4
  comformable, 8
  diagonal, 13–14
  difference of, 5, 16, 17, 21
  elementary row transformation, 44
  equal, 3
  multiplication of, 7
    associative property for, 10
    distributive properties for, 9, 10
  notation for, 2, 4, 5, 14, 16, 19
  Pauli, 12
  product of, 7, 16, 17. 21
  reflection, 58
  similar, 93
  sum of, 4, 16, 17, 21
Matrix, 2
  additive inverse of a, 5
  anti-symmetric, 15
  augmented, 46
  characteristic equation of a, 83
  characteristic function of a, 83
  characteristic values of a, 86
  characteristic vectors of a, 86
  classical canonical form of a, 93
  of coefficients, 6, 46
  of cofactors, 32
  column, 8
  complex, 19
  of a conic section, 104
  conjugate of a, 19
  determinant of a, 22–24
    value of a, 22–27
  diagonal, 13, 61
  diagonalization of a, 93

Matrix (*cont.*):
  echelon form of a, 43, 47
  eigenvalues of a, 85
  eigenvector of a, 86
  element of a, 2
  Hermitian, 19
    eigenvalues of a, 90
    eigenvectors of a, 91
  identity, 14
  latent values of a, 86
  latent vectors of a, 86
  modal, 93
  multiplication, 7
  multiplicative inverse of a, 29, 32, 45,
    99
    left, 28–29
    right, 29
  nonsingular, 32
  null, 4
  order of a, 2
  orthogonal, 68
    improper, 69
    proper, 69
  postmultiplication of a, 8
  premultiplication of a, 8
  proper values of a, 86
  proper vectors of a, 86
  rank of a, 42
  real, 2
  reflection, 58
  rotation, 55
  row, 8
  r-rowed minors of a, 42
  scalar, 13, 61
  scalar multiple of a, 5
  singular, 32
  skew-Hermitian, 20
  skew-symmetric, 15, 16
  square, 2
    order of a, 2
  symmetric, 15, 16, 96
    eigenvalues of a, 91
    eigenvectors of a, 92
  trace of a, 84, 85
  translation, 73
  transpose of a, 16, 19
  unit, 14
  upper triangular, 28
  zero, 4
Minor:
  of an element, 26
  of a matrix, 42

Modal matrix, 93
Multiplication:
  of matrices, 7
    associative property for, 10
    distributive properties for, 9, 10
  of quaternions, 39
Multiplicative inverse of a matrix, 29, 32,
    45, 99
  left, 28–29
  right, 29

Noncentral conic, 105
Nonhomogeneous coordinates, 72
Nonsingular linear homogeneous trans-
    formation, 66, 67
Nonsingular matrix, 32
Null matrix, 4
Null space, 68

One-to-one mapping, 52, 66
Operator, 7, 55
Order, 2
Orthogonal matrix, 68
  improper, 69
  proper, 69
Orthogonal transformation, 96
Orthogonal vectors, 55, 69, 91, 92

Pauli matrices, 12
Point transformation, 53
Postmultiplication of a matrix, 8
Premultiplication of a matrix, 8
Principal Axes Theorem, 96
Principal diagonal, 13
Product:
  transformation, 56–57
  of two matrices, 7, 16, 17, 21
  of two quaternions, 39
Projections of the plane, 65
Proper:
  conics, 110
  orthogonal matrix, 69
  values, 86; see also Eigenvalues
  vectors, 86; see also Eigenvectors

Quadratic form, 101
Quaternion, 39

Rank:
  of a conic, 109
  of a matrix, 42
Real matrix, 2
Rectangular array, 1, 2
Reflection matrices, 58
Reflections of the plane, 58, 76–77, 79–80
Right-hand distributive property, 10
Right multiplicative inverse, 29
Rigid motion transformations, 79–80
Ring, 35–36
Rings, isomorphic, 36
Rotation:
  matrix, 55
  of the plane, 54, 76–77, 79–80
Row:
  index, 2
  matrix, 8
  transformations, 43
  vector, 8
r-rowed minors, 42

Scalar matrix, 13, 61
Scalar multiple of a matrix, 5
Shear:
  parallel to the x-axis, 63
  parallel to the y-axis, 65
Similarity transformation, 93
Similar matrices, 93
Single-valued mapping, 51–52, 66
  into, 51–52
  inverse, 52
  one-to-one, 52
  onto, 52
Singular linear homogeneous transforma-
    tion, 122
Singular matrix, 32
Skew-Hermitian matrix, 20
Skew-symmetric matrix, 15, 16
Square:
  matrix, 2
    order of a, 2
  submatrices, 42
Sum of two matrices, 4, 16, 17, 21
Symmetric matrix, 15, 16, 96
  eigenvalues of a, 91
  eigenvectors of a, 92
Systems of linear equations, 6, 46
  consistent, 47, 50
  homogeneous, 49–50
  inconsistent, 47

Trace, 84, 85
Transformation; *see also* Transformations
    of the plane:
  diagonalization, 93
  identity, 56
  invariant under a, 58, 69–70, 80, 81,
    109–110
  orthogonal, 96
  point, 53
  product, 56–57
  similarity, 93
Transformations, elementary row, 43
  matrices, 44
Transformations of the plane, 53
  dilation, 61
  general linear, 71, 80, 81
  linear homogeneous, 66, 67, 80
    nonsingular, 66, 67
    singular, 122
  magnification, 61
  null space of, 68
  projections, 65
  reflections, 58, 76–77, 79–80
  rigid motion, 79–80
  rotation, 54, 76–77, 79–80
  shear:
    parallel to the x-axis, 63

Transformations (*cont.*):
  shear (*cont.*):
    parallel to the y-axis, 65
  translation, 73, 79–80
Translation:
  matrix, 73
  of the plane, 73, 79–80
Transpose of a matrix, 16, 19
Transposition, 16

Unit matrix, 14
Upper triangular matrix, 28

Vector:
  column, 8
  row, 8
Vectors:
  dot product of, 8
  inner product of, 8
  orthogonal, 55, 69, 91, 92
  scalar product of, 8, 69–70
Vector spaces, invariant, 88

Zero:
  divisors, 11
  matrix, 4